THE ENERGY WAVE FIELD

BY ROBERT M. MATTER

DORRANCE
PUBLISHING CO
EST. 1920
PITTSBURGH, PENNSYLVANIA 15238

Dorrance Publishing Co
585 Alpha Drive
Suite 103
Pittsburgh, PA 15238
Visit our website at *www.dorrancebookstore.com*

ISBN: 978-1-6470-2456-7
eISBN: 978-1-6470-2705-6

For Sara

FOREWORD

It was one of those "eureka" moments when many years ago, I noticed a particular interference moiré pattern of oppositely pulsating pairs of what were essentially 2-D projections of 3-D spheres, appropriately labeled as *spheric wave-forms*, in contrast to *linear wave-trains*. A second interference pattern of a linear wave train clearly contradicted what had been interpreted from the Compton experiment, namely, Einstein's assertion that light traveled as "quantum point-particles." Having no qualification in the realm of physics or quantum theory, I'd nevertheless been reading and pondering some of the assumptions associated with gravity, quantum theory, and mathematical rules applied to the physical world, in particular, the "wave-particle duality" and the relationship of energy as conserved, to matter. The secondary moiré patterns representing wave interferences were of longer wavelengths that Einstein-Compton maintained could *not be* waves and had to be particles (shown in figure 17A and B in text). Einstein's assertions appeared particularly questionable, but when seeing the oppositely pulsating pairs of standing spheric-waves (figure 35 A and B, p.TBD), I immediately realized the mistake of the wave-particle dualism, and linked those pairs of opposite pulsings to Pauli's "exclusion principle" and/or to the "particle / anti-particle" pairs of Dirac.

I also pondered the meaning of "Conservation of Energy", realizing that Energy itself is invisible and immeasurable, and inferred only from its measurable effects in systems' behaviors. Being neither created nor destructible, Energy is therefore timeless (as Emmy Noether proved mathematically), and must necessarily have primacy over such secondary effects as the spheric "matter-waves". It was realized too that "entropy" as currently defined could have no relationship whatsoever to Energy. Only the wave interference configurations as *effects* of Energy wave interferences are subject to entropy; Energy itself is not.

Louis deBroglie raised the question whether both matter and energy behaved like "particles and waves", but the developement of quantum theory seems to have emphasized the "particle" over the wave (thanks to Einstein's insistent arguments). Citing a 2007 discovery in photosynthesis concerning "long-term quantum coherence" and "harmonic resonance" of frequencies between plants' cell-molecules and the sun's radiant energy, the "quantum" is suggested to be a "harmonic resonance" ratio between emission and absorption of electro-magnetic radiant energy with matter—as Planck consistently maintained—and which are both considered to be waves in this presentation (as deBroglie and Schrödinger believed, and not at all as "particles" of either energy or matter). "Atoms" and "molecules" are seen as secondary "wave complexes" of spheric wave forms in interference with radiant energy waves, the quantum being a harmonic resonance ratio of their frequency, wavelength and intensity interactions.

Following a brief history and a primer on wave terminology and interactions, this work focuses on wave interferences, the primacy of Energy over matter, and of wave over particle, with implications and directions for further research. Following the section on basic quantum theory, the paper turns to cosmological assumptions, questions and implications, ending with a more detailed and comprehensive look at the Universe as an Energy Wave Field.

April 5, 2019

TABLE OF CONTENTS

    Background: Energy, particle-wave duality, and cosmology

    Waves, a primer

    Electro-magnetic wave spectrum

    Wave types and terminologies

    Wave interactions—reflection, refraction and diffraction

    Wave interferences; superposition

    rectilinear/curvilinear and spheric waves

    Radiation/particle; Huygens; Hydrogen's electron waves

    Photo-electric effect; strings

    Compton effect; spiral propagation

    Laser; Photosynthesis

    Field

    Real and Imaginary Number Systems

LIST OF ILLUSTRATIONS

PREFACE

HISTORICAL BACKGROUND OF ENERGY, THE "WAVE-PARTICLE" DICHOTOMY, AND THE "EXPANDING UNIVERSE."

The word *energy* is derived from the *Greek en + ergon* which literally translates as "in" or "at work", which means action without any actor. Energy is pure *self-acting action* with *no thing or entity* that is acting. The idea of energeia as a *vis viva*, or "vital force" first appears during the 4th century BCE in the *Nicomachean Ethics*. In 1918, *Dr. Emmy Noether* published her paper that mathematically *proved* the *Law of Conservation of Energy*, that because the laws of physics do not distinguish between different moments of time, *entropy*—as the "arrow of time".-*can have no relationship whatsoever to the greater concept of Energy conserved.*

The Law of the *Conservation of Energy* means that Energy is indestructible and therfore timeless. This paper is meant to serve as a corrective to the current path of physics and of cosmological theory. There are therefore three important themes throughout this paper. The first is the explicit emphasis that *Energy is primary, and that gravity, radiant energy and matter are three of the secondary effects of a Universal Field of Energy Wave interferences,* The second is an adjunct to the first, that what I call 'specious quasi-particles" are actually spheric *wave-forms*; as energy is primary over matter, so this paper will demonstrate why *waves are primary over particles.* It is a complete and necessary reversal of what is now called "particle physics", which is an outcome and a branch of *quantum physics* that focuses on material "particle-things". This will be made clear in the text and illustrations, especially when the intersecting waves can be seen *in motion*, which unfortunately cannot be shown in this paper; I can only show *directions of motion* with arrows where applicable. The third is that the entire universe is nothing but Energy that functions in and as different wave configurations.

Perception of the duality of "particle" and "wave" goes back for millennia to *Pyhthagoras* and his followers during the 6th century BCE, and to the paradoxes of Zeno the Eleatic during the 3rd to 2nd centuries BCE. From *Plato* to *Aristotle* to *Archimedes*, the dualism of the continuous and the infinitely divisible discrete i.e. "infinitesimal" surged as topics for discussion until the dualistic conflict between the mathematicians and the Jesuits became political during the 1500's OCE.[1] *Isaac Newton* (17th to 18th centuries) believed that a beam of light was made up of a stream of "discrete corpuscles". During that same time period, the mathematician *Christian Huygen's* experiments with optics convinced him that light consisted of waves. *Reflection, refraction* and *diffraction* were easily explained by a wave theory of light, but Newton thought there was some problem with the refraction of straight rectilinear waves, and asserted his own corpuscular theory. In figures 10 A and B, it is clear that there is no "problem" with the refraction of rectilinear waves. In the 17th centuries, *Gottfried Wilhelm Leibniz* defined *energeia* as a "living force", and first offered the physics of a material object as its *mass* (m) *times the square of its velocity* (ie. $e = mv^2$). The weight of Newton's reputation led to the dominance of his theory until in 1803, *Thomas Young* first performed what is known as *"the two-slit" diffraction experiment* that clearly demonstrated the wave theory of light. In 1807, Young published that *"the product of the mass of a body into the square of its velocity may properly be termed its energy"* (Note the nearly identical form of Liebniz and Young's relationship of e to mv^2 and Einstein's *E* to mc^2 where "E" is capitalized and "v" is replaced by "c" as Light's absolute velocity in vacuum.) In 1881, *William Thomson*—also known as "Lord Kelvin"—stated that:

> *"The very name 'energy" first used in its present sense by Dr. Thomas Young, has only come into use after its definition had raised it from a mere formula to the position it now holds of a principle pervading all nature and guiding the investigator in the fields of science." [2]*

Around 1860, *James Clerk Maxwell's* four equations forever united electric and magnetic forces as two aspects of the same natural force called *Electro-Magnetism* (EM), a *radiant energy* that travelled as waves in vacuum at the speed of light (this was later confirmed, and in 1887, corroborated by *Heinrich Hertz*). During the 1800's, the wave theory of light was upheld by experiment until—as will be shown in the text—*Albert Einstein* in the early 1900's reversed that practically by decree. The *quantum* had its theoretical origins in the earlier 1900's when *Max Planck*, wrestling with classical Newtonian equations that were contradictory to what actually happened in nature and in experiment, was drawn to the conclusion that *emitted energy interacted with matter as bits or chunks* whose *quantity* was determined by what was absorbed or emitted from hypothetical "oscillators".[3] Planck had written that he didn't believe that energy itself was quantized, but that quanta appeared *only* when energy interacted with matter. Later, at the September 1909 *Salzburg Conference* (attended also by Einstein), Planck emphatically restated his firm belief that energy *appeared* as quantized *only when interacting with matter*.[4] Hertz had been experimenting with the *photo-electric effect,* which led to Einstein's photo-electric experiment where he insisted that energy itself was emitted as particles.

If Energy is indeed a *"principle pervading all nature"* and *as conserved* is indestructible, timeless and eternal, then as self-acting action, it must be the fundamental ground of all mass and gravitation, all radiation and matter, of the entire universe and us in it. It is, therefore, *omnipresent*. It is pure "doing with no doer", which means it is also *omnipotent*. That is what the "energy conserved" implies. As "all pervasive", the question that comes to mind is "What would be the *difference* between "Energy waves" and "Gravity waves" when both are invisible and only *inferred* from observable perturbations of material objects?

As Energy is universal and eternal, alternatives to the current "standard cosmological model" and the "accelerating rate of expansion of the universe" will be examined. From the 16th through the 18th centuries, it was thought that the mechanics of matter could explain the world, so there was a kind of "taboo" on anything suggestive of religion or *incorporeality*, and attention was focused on observable "matter". But in spite of the modern concept of conservation of energy, this "physicalist" focus continues to this day. Einstein stated that all matter "contained" energy, the standard view being that "atoms" *have kinetic energy* (Ek) when in motion, and *potential energy* (Ep) as energy stored at position, plus a *rest energy* as a "ground state" when an atom or body is at rest. But "rest energy" *is still the Ek and Ep* as the measure of an atom's proton and neutron *oscillations*, and the *spinning* and *orbiting* "energy of electrons"[5], and this is what Einstein meant. But according to the EWF hypothesis, Einstein's statement is incorrect; *matter is itself nothing but Energy functioning in and as an interference moiré wave form, configuration or pattern.* Since form-pattern can be destroyed and energy cannot be destroyed, *entropy can apply only* to form and pattern which is subject to change and destruction, but the *Energy that functions* as form and pattern is itself indestructible and timeless.

The concept of *field* is *a measure of some physical attribute* that is spread out over every point in space-time. Energy Waves constitute such a Field (EWF) in which the unmanifest can manifest by interferences that produce matter and radiant energy waves. As unmanifest, the EWF is immeasurable; it is only when self-interfering waves manifest as secondary effects (such a matter) that measure becomes possible. These interferences of two or more waves represent no dualism because the "self-acting action" of Energy is a Möbius-like twisting and bending action whose self-interferences manifest as moiré configurations or forms. The Real Number System and its arithmetic definitions, laws and rules is examined along with "the calculus" as related to quantum theory, including the implicit assumption that the real world of the universe is *intelligible*, and what is deemed "intelligible" has always indicated *intelligence*. Energy is posited as a field of vibration waves whose self-interferences manifest as our familiar "material" world. The *Energy Wave Field* greatly simplifies the hodge-podge of *ad hoc* adjustments that patch and prop up the collapsing "standard model", and has potential for unification of what today is contradictory and incomplete, despite many successful predictions.

My admittedly personal bias in cosmology is based on the history of an *assumed* connection between the then recently discovered *Doppler effect* and the almost contemporary observation of deep

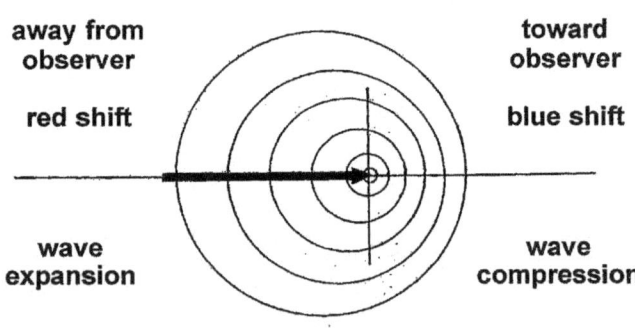

Preface 1 - Doppler effect

space *red-shifts which have set a precedent* that, when repeated over a period of almost a hundred years, has become established as "quasi-fact", like a judge's court-ruling which sets a precedent that becomes the "*color of law*" in future trials. Shortly after the discovery in 1842 by *Christian Andreas Doppler* of a phenomenon of changing sound pitch, (See Preface-1) *Armand-Hippolyte-Louis Fizeau* in 1848 was the first to assume a connection between the observed red-shift of light from stars and the *"Doppler effect"*. This assumed link was followed by William Huggins in 1868, "confirmed" in 1871 and again in 1901; the assumed link was repeated in 1912,1923, and in 1931. That first assumption in 1848 set a precedent for the "runaway expansion" theory of the universe. There is one experiment that seems to uphold that assumed connection; recording the red-shifts of several deep-space objects over an interval of time, the changing red-shifts for each object are proportionate to their estimated distances. But there are other hypotheses that can also explain the same relationship of red-shift to distance (discussed in the "Cosmology" section).[6]

Another point is the assumption that *entropy* applies to the Universe as a whole. As stated above, if all Universe in *any* of its "changing states" is timeless Energy, entropy can *only* apply to the *form-configurations* of wave interferences. As explained in the text, the projected "finalities" such as "thermal equilibrium" of an entropic universe lead to contradictions.

Even in science, personal belief and bias, the momentum of past authority, or the weight of currently favored doctrine all contribute to the selection of hypotheses to be tested. In addition, it takes time and grant-funding to actually perform a fraction of experiments for the many hypotheses put forth for testing. There is also peer pressure and politics, and the temptation for potentially lucrative commercial or military applications. Objective science is not without subjective-psychological, philosophical and personal valuations, and all science is grounded in some metaphysical belief, the current materialistic-based "reality" being no exception. Since all knowledge—scientific or otherwise—is cumulative, incomplete and unending, it should never be mistaken for fact, much less for Truth.

The sciences are usually divided into theoretical and experimental branches, but there is also discovery, which is often made serendipity by amateurs or other non-professionals, or by "intuitive thought experiments" such as Einstein's "relativity", or Planck's 'quantum'. In any case, an actual theory follows only *after* the experimental scientists design the apparatus to test the hypothesis for confirmation or nullification. In the text that follows I will sometimes abbreviate quantum theory as "QT", big-bang as "BB", Electro-Magnetism as "EM" and other acronyms as commonly used in physics and cosmology. Capitalized "Energy" refers to the unmanifest immeasurable Energy prior to or during self-interference, while lower-case "energy" is used for measurable energy as it functions as form or system. "Light" when capitalized refers to the whole EM energy spectrum, but lower-case "light" refers only to that small range of EM energy to which our retinas are sensitive. "Universe" and "universe" are used inter-changeably, depending on context. There are five main points emphasized in this paper:

1) Energy as conserved is indestructible, and therefore timeless and eternal;
2) Energy is self-acting action; *there is no actor*;
3) The 2-D model of a 3-D spheric wave in self-interference yields both secondary pulsing spheric wave-forms as well as linear wave-trains; gravity, EM radiation and matter are all direct *effects* of an otherwise unmanifest Energy Wave Field
4) An *intelligible* Universe signifies *intelligence*;
5) The *Energy Wave Field* must ultimately be referred to as The *Mind-Energy Wave Field.*

I hope the reader will acquire and use transparencies as indicated in this paper, or better yet, use on-screen computer graphics to see first-hand the *motions* of wave interference *moiré* configurations that are so arduously attempted to be described here. I can only show the directions of movement with arrows in this paper. Actually *seeing* the directional and frequency changes of the *secondary wave effects—especially the oppositely pulsing pairs* of secondary "closed and standing" *spheric matter-wave forms*—is enlightening. In the static figures that follow, black lines are the "+" *wave crests*, and the white spaces are the "—" *wave troughs.* The only exception to that is *Figure 13, where the black represents troughs and the white as tight-wave crests that interfere together to cancel out into black darkness.* In actuality, the wave crests will be white and the troughs black (ie., reversed from the printed graphics). All these terms will be clarified in the text and figures that follow the introduction to the nomenclature, basic types and patterns of waves and their interference configurations that make up what I refer to as *The Energy Wave Field (EWF).* Those schooled in wave mechanics should become familiar with the following terminologies that sometimes differ from those in standard elementary physics textbooks.

INTRODUCTION

WAVES A PRIMER

Some waves are formed as effects of natural disturbances, like wind on the surface of an ocean or lake, or sand-dunes, or the waves of a vibrating guitar string, or compression waves in a pipe-organ. But all waves are created by some disturbance, and can be conceived as vibrations travelling away from the source of disturbance. All physical waves travel through some *medium*, like air, water or steel; but in the case of *electro-magnetic* waves, the source of the disturbance is said to be a moving electrical charge or some magnetic oscillation. There being no "luminiferous aether" pervading the universe as a medium through which electro-magnetic waves propagate, a question posed is "Well then, what waves?" In Maxwell's standard wave theory, the "medium" *is* the EM field itself. But then the question is "In what does the charge *move*?" The answer, I suppose, would be "in space-time" which, in this paper *is* The Energy Wave Field.

A *periodic* wave *set*—*sometimes called wave packet*—is any group of wave-vibrations travelling or propagating at the same rate in the same direction. These can be *open wave-trains* of *radiant energy waves* such as light or heat or x-rays, each type traveling at its own *frequency and amplitude and in the same direction* through space-time from some disturbance; or as *closed and standing concentric or spheric wave-forms* that usually come in oppositely pulsing pairs as "matter-waves". Any set of waves that interact with other wave sets, including matter-waves, are said to be in *interference*, which seems almost always to produce a *secondary wave-effect of a lower frequency and longer wavelength*. These effects are called *moiré patterns or configurations*. A *wave front* is any wave that initially propagates from a disturbance, or any wave that happens to be incident upon or into another *wave-complex* at any particular moment. "Wave-front" is a term relative to a specific observation of a particular wave system. **Figure 1** shows a train of electro-magnetic waves with a theoretical center line of directional propagation dividing upper *crests* valued as "plus", from lower *troughs* valued as "minus". *Amplitude* (*a*) is the height of crest *above* the center line, or the depth of the trough *below* the center line, and *intensity* (*i*) is the *amplitude squared* ($i = a^2$); *Wavelength* (L for the Greek *Lambda*) is the distance from one crest (or trough) to the next adjacent crest (or trough); Wave frequency (*f*) is the number of *cycles per second*, which are also named *Hertz* (*Hz*), and a "cycle" is more or less equivalent to one wavelength (although as measured at the dots on the "center-line"); wavelength and frequency are *inversely proportional* to each other, so that as wavelength shortens or lengthens, frequency becomes respectively higher-faster or lower-slower. *Frequency and amplitude are both measures of the energy function* of a wave-set, and each has a different *effect* as described in the text. Waves are *periodic* as a specific *interval of time* between two or more successive waves.

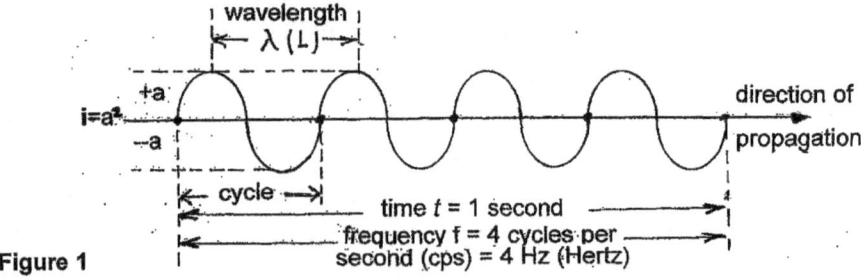

Figure 1

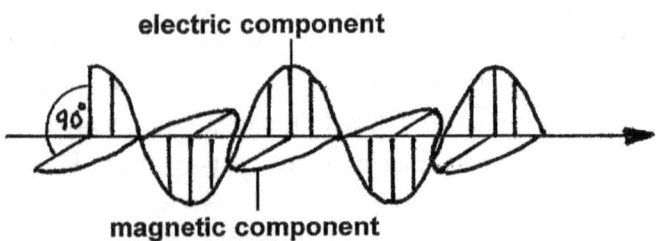

electric component

magnetic component

Figure 2 - Electro-Magnetic Wave

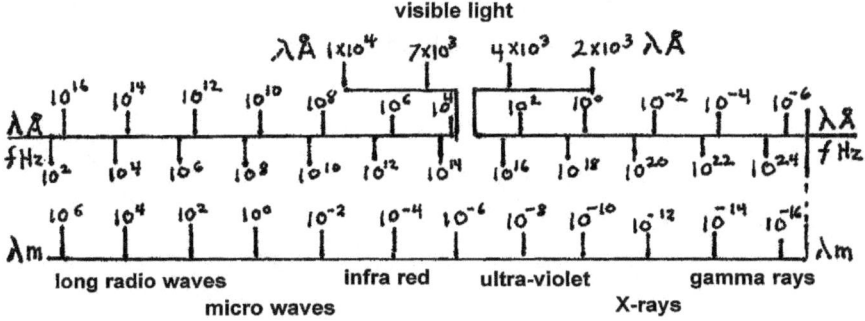

Figure 3 - Electro-Magnetic Spectrum

Electro-magnetic (EM) waves have an assumed electric and magnetic component which are thought to be perpendicular to each other as well as to the line of propagation, both traveling in vacuum at the constant speed of light (c). Both electric and magnetic waves illustrated in **Figure 2** are presented as *transverse waves* that travel *in phase* at the same amplitude, frequency, and wavelength. All EM waves are periodic wave-trains, but in Figures 7D and 8 below, Energy Waves in *interference* also appear as wave-*forms*. Clerk Maxwell's theory combines all EM energy waves into one spectrum with long wavelenghs at low frequencies invisible to human eyes (at left in **Figure 3**), such as radio and infra-red waves, through a very narrow section that is visible to human retinas (center), to more invisible of ever shorter wavelengths and higher frequencies, such as ultra-violet, x-ray and gamma rays at the opposite end (to the right in Figure 3). As you can see from radio and television, not to mention wireless cell-phones and wi-fi, there is much more to EM energy than meets the eye, and what is *visible* is a miniscule slice of the total EM spectrum. Wavelengths at the low end can be measured in feet, yards or even miles, while wavelengths at the high end are measured in micrometers, nanometers (nm) or Angstrom units times ten to the nth "+" or "—" power, depending on which part of the spectrum is measured and in what units. Energy Waves are partially modeled after EM energy waves that are focused in this section.

There are four main types of waves: One is the *transverse wave* as seen in Figure 4, which can be like a rope being flipped (**Figure 4 A**), or as most "transverse waves" are, a *cross-section* of a wave-train like the edge of a corrugated roof (**Figure 4 B**). When physicists think of waves, they usually picture a *transverse* *sine-wave* instead of graphically depicted *wave-trains*, and the *moiré pattern effects* from interferences seem not to be considered.

Figure 4A   Transverse Waves   **Figure 4B**

The second type is the *rippling wave-train*, either as *rectilinear* (**Figure 5 A**) or *curvilinear* (**Figure 5 B**). Rectilinear or "straight" wave trains can be formed in nature, but a curvilinear may *appear* straight at the distant peripheral radius (at far right in Figure 5 B). Curvilinear trains radiate from spheric bodies as "shells" of crests and troughs. These, with the "concentrics" and "spherics" in Figures 7 and 8 are those that produce the most interesting interference patters as seen in Figures 15, 16, and 17.

Figure 5A   Rectilinear and curvilinear   Figure 5B

A third wave type is the *compression wave*, of which there are two types; one is *longitudinal*, like a spring in **Figure 6 A,** or as air compression in a tube, like a flute or organ-pipe. Other sound waves *radiate spherically* outward in air from some center of disturbance (**Figure 6 B**).[7] Technically, 6-B is a concentric radial wave, like 7 A and B below; but since "sound" is a compression of air or fluid *molecules*, it can also be classed as 6-B "compression wave", but is *not* longitudinal. However, when sound compression is restricted to a solid like an iron bar, or a tube as in figure 6A, then it is propagated longitudinally. In any medium, sound waves are the compression and rarefaction of the molecules of the medium, whether longitudinally (6A) or radiating spherically as in 6B.

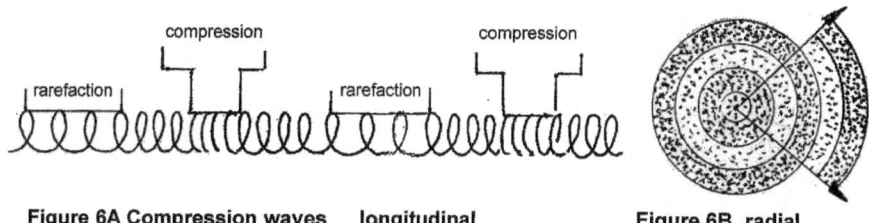

Figure 6A Compression waves   longitudinal   **Figure 6B radial**

If you watch raindrops making ripples on water, you can see how the waves radiate outward in circles from a center of disturbance—like a pebble dropped in a pond **(Figure 7 A,B)** or the light from a burning or exploding spherical star in the cosmos. There are two versions of the fourth kind of wave: *concentric waveforms* have evenly spaced waves (**Figure 7 C**), and *Spheric wave-form* that seem to be more and more compressed as the radius increases, (**Figure 7D**) but this appearance is due to the 3-D *spheric* form of this particular wave when projected onto a 2-D surface as represented in **Figure 8.**[8] The actual frequency of a 3-D spheric wave can be evenly spaced pulsations, but when projected onto a 2-D plane-surface, appears as though the frequency and wavelength change at the farthest perimeter. If they actually *did* change in wavelength and frequency in 3-D, it would have to mean that *time as duration had also changed*, either via relativity, or by gravity, or via refraction into a different medium (see below, Figure 10) or as some unknown cause. But as a 3-D spheric wave, the frequencies do not change, even if they appear to do so when projected in 2-D.

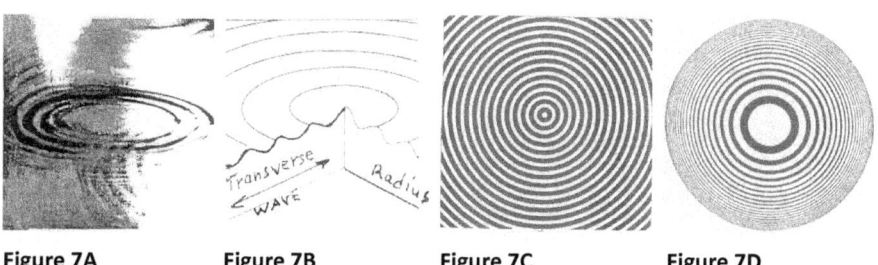

Figure 7A   **Figure 7B**   **Figure 7C**   **Figure 7D**

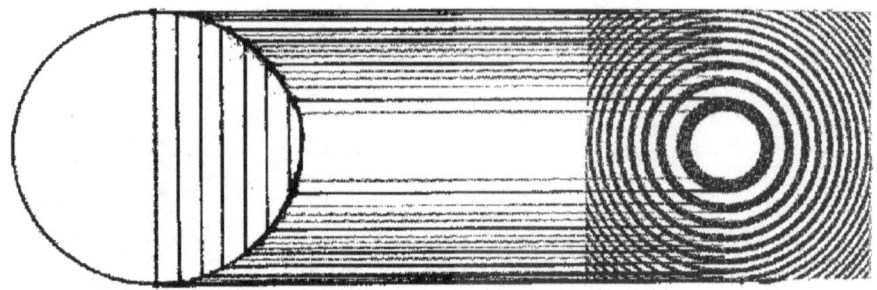

Figure 8 Increment slices projected right as a spheric wave

Reflection is a "bouncing" of Light from some non-absorbing reflective material, and the rule is that the *angle of incidence with which light strikes a surface is equal to its angle of reflection,* (It is complicated having to draw all the waves, so most illustrations are always simplified to a single line-ray for both incidence and reflection as in **Figure 9A and B**. However, the waves are also drawn in Figures 9 A and B as a reminder.) **Figure 9 C** shows the reflection of one single - pulse wave. **Figure 9D** is a photo of periodic concentric waves in a water-tank "bent back" or reflected toward its vibrating source.

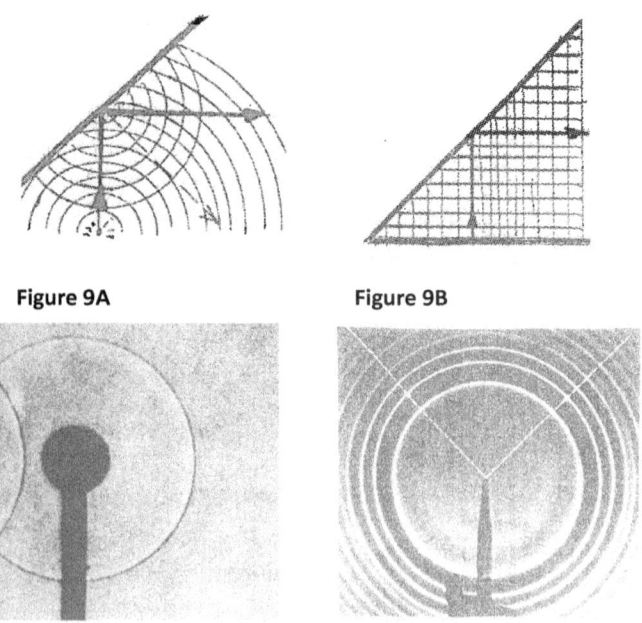

Figure 9A   Figure 9B

Figure 9C   Figure 9D

Refraction occurs when light travels from one medium's density into that of a different density, and the angle at which the wave-train is bent is proportional to the two densities, to the angle of refraction and to the wavelengths. When white light through a prism is bent, each color is individually refracted by wavelength at different angles into separate bands of the visible red, yellow, blue and violet spectrum colors. Considering that there is a fixed ratio between any specific frequency and its correlate wavelength in the EM spectrum, stating that as speed changes, wavelength changes but frequency remains constant seems incorrect. As speed changes between different densities, *both frequency and wavelength must change together* to maintain that ratio regardless of speed. The standard $c=Lf$ indicates that when "c" becomes a variable speed (*S, or v*) both L and f must change. The equation $v=Lf$ remains, but the interpretation changes so that *all three terms must change together*. Refraction is an *acceleration* of light waves.[9]

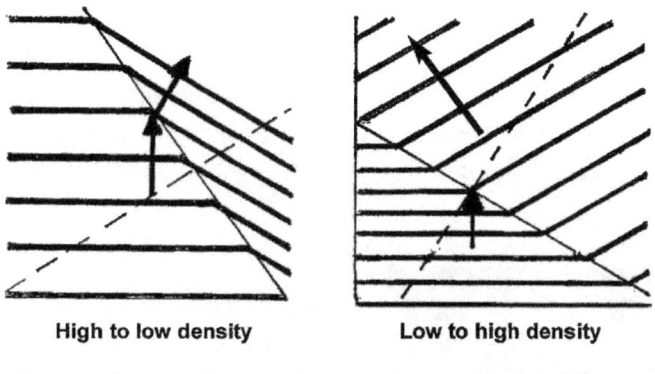

High to low density   **Low to high density**

Figure 10A   Refracting Waves   Figure 10B

Diffraction: In the late 1700's, *Thomas Young* shined a single beam of light through a single slit which diffracted around the slit's edges into a semi-circular wave-train that then passed simultaneously through two more slits making two semi-circular wave-trains that are projected onto a screen as shown in **Figure 11 A**. The pattern of alternating dark and light "bars" was due to the *interferences* of wave crests and troughs, and proved that light was a wave phenomenon. As well as diffracting through small apertures, light also appears to diffract around edges of objects. For example, sunlight behind a pole or tree appears to diffract around the object's edges (**Figure 11 B**), indicating that when light meets the *edges* of a material object, there is some kind of "drag" on the wave-front as it passes around or through a slit or aperture of a material obstacle. However, 11 B might also be just an optical "glare-effect" of the eye's lens.[10] When it was reported that a single particle—either a photon or an electron had been "diffracted" in a two-slit experiment, and that a single particle had instantaneously gone through both slits, it was asked "How is that possible? How can one single particle go through two slits at the same time"? The standard quantum explanation as established by *Louis de-Broglie* is that the electron and the photon each behave as "particle and wave", depending on the experiment. In the photo-electric effect, both "photon" and "electron" behave as particles, but in the two-slit experiment, they both act like waves. Despite the "wave-particle" duality in Q-theory, Maxwell's *wave* equations had been "quantized" when re-worked by *Paul Dirac*, and EM waves became treated as "photon-particles". But it turns out that *no single photon has ever been released or passed through any slits*. The hypothetical "single photon" of the double slit diffraction experiment had never been executed, nor could it ever have been possible to execute. In quantum theory any intrusive action of *any* kind—controlling the direction, measuring, even *releasing* a "single photon"—would make such an experiment impossible. It was only a "thought experiment" and not a physical actuality which makes it not even a hypothesis since it could never be tested.[11] It makes one question even the diffraction of "a single electron".

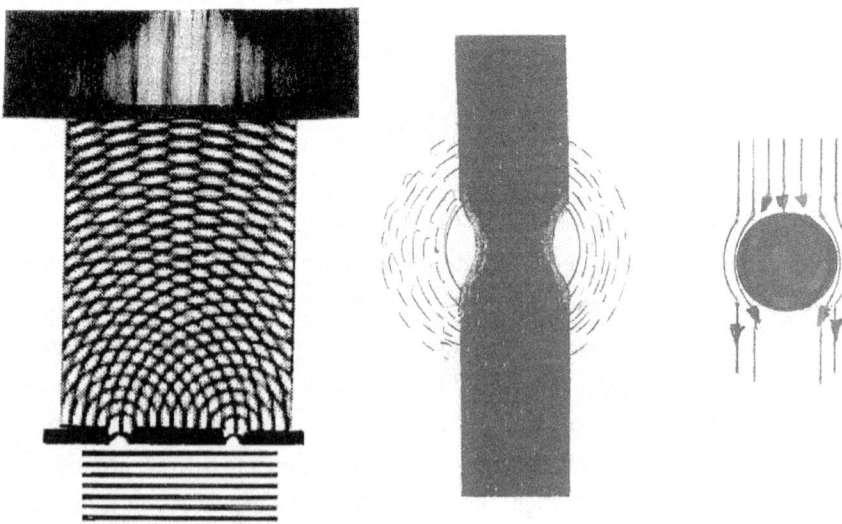

Figure 11A Wave Diffraction   **Figure 11B**

In the following two equations, the *Planck-Einstein formula, Energy (E) = Planck's constant (h) times frequency (f)*, and in the *Louis de Broglie formula, momentum (p) = h divided by wavelength (L)*, *each* equation contains both a "*quantum particle*" part (*h*), and a *wave property*" part (*f and L)*, illustrating the quantum "particle-wave duality". Despite the visual evidence of wave interference configurations, and despite the indisputable evidence by deBroglie that "particles" display unmistakable *wave* characteristics, in contemporary Quantum theory, particle physics has all but replaced the wave aspect of quantum mechanics. As shown below in **Figure 12 A**, if an "electron" is seen as an *actual standing* 3-D spheric wave, it can and does easily diffract and pass through both slits at the same time. But if electron is a *particle* **(Figure 2 B)**, going through two slits at once is a mystery (although a stream of them could diffract through two slits). Keep in mind that the slitted *barriers themselves* are also complex matter-wave structures that computer graphics should be able to show. In essence, *the electron or photon wave is in interference with the matter-wave-complex of the "material barrier"*. The matter and EM energy-waves, being spheric, only *appear* as "particles", but are only "quasi-particles".

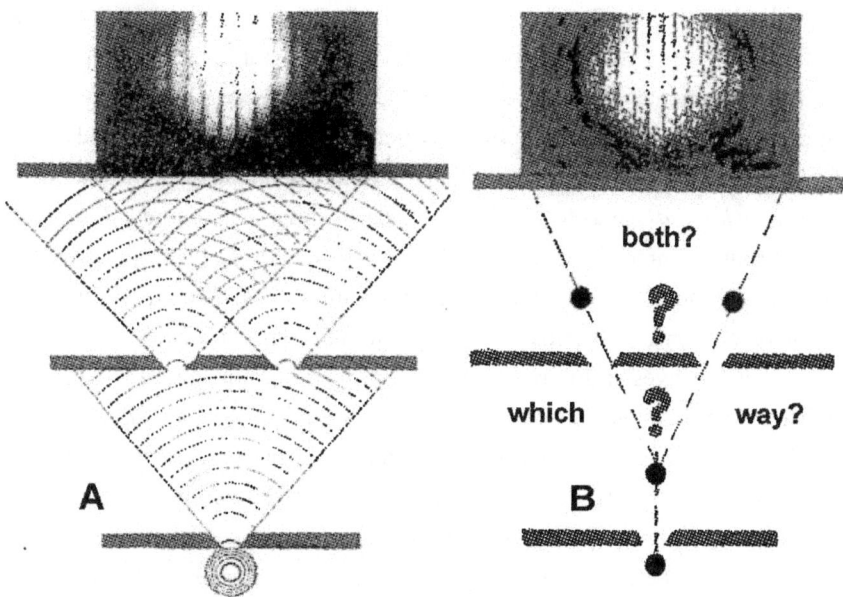

Figure 12--Diffraction of A--wave; B--particle

WAVE INTERFERENCES[12]

Note that there is a crucial difference between real waves and the 2-Dimensional wave patterns shown in these pages. The *movement of transparencies* is completely reversed from what actually happens. When moving transparent overlays to observe the motions and pulsings of secondary wave effects, *the printed spacings of lines as wavelength are fixed, but the overlays can be moved at variable speeds that would change the frequencies.* In actuality, *all real waves travel at speed of light-c, so the rate of propagation is more-or-less fixed, while the wavelength relative to frequency is the variable spacing of wave crests per unit of time.* Besides the spacing of crests per unit of time, any pulsings of secondary waves especially those of paired spheric matter-waves are equivalent to a single oscillation, with n-pulsings-per-second being the frequency. But each pulse is *not* a "quantum", and especially *not* a "particle"; a pulse is just a single wave-cycle. This is also why computer graphic simulation is a much better medium to observe and measure wave-train and wave-form motions. The speed of waves can be fixed while allowing manipulation of direction, frequency and wavelength. As presented here in a very limited and limiting medium, interference effects will hopefully stimulate the use of computer. There is one more crucial difference: the graphic black and white would be reversed in real waves.

There are two basic types of wave interferences, but both kinds are derived from the fundamental *unmanifest*, self-acting Energy Waves simultaneously pulsating *out from*, and *into* an emptiness which is the EWF itself. As fundamental Energy, this single pulsating wave would be timeless and have an infinite radius, *"turning inside-out"* so as to *self-interfere.* This primary and perpetual self-interference produces three manifestations: 1) *Initial gravitational curvature* as the fundamental geometry or architecture of space-time;[13] 2) Free-travelling open *rectilinear* or *curvilinear wave-trains* that can interfere with each other or with spheric matter-waves at different angles, regardless of their specific frequencies; 3) closed, standing concentric or *spheric matter-waves, often manifesting as oppositely pulsing pairs* as though echoing in time and space the timeless inside-out of the primary Energy Waves. Both 2 and 3 above are known wave types, but all three manifestations are inter-dependently interactive, since they are all just different aspects of the Universal Energy Field. It was the third type that caught my attention when I saw the *oppositely pulsing pairs* that I interpreted either as opposite "charge" or opposite "spin." "Charge" is essentially the *direction of current,* so if translated to anything, it would most likely be "charge" All waves, including EM waves, that interact with other waves or wave complexes might affect the *initial* wave-trains' directions, amplitudes and the frequency/wavelengths, and will certainly produce new secondary wave-sets as *effects* of the initial set interferences. In this text, *initial waves* are two or more wave-sets that interfere with each other, effecting *secondary waves as moiré configurations*. These latter terms—*"initial"* and *"secondary"*—are applicable mainly to *particular interference situations (ie. >0 and <45 degree intersections)*.

The simplest interference is when two waves of the same frequency and amplitude (**Figure 13 A and B**) are traveling *"in-line" in the same direction*, but are *out of phase* by a half-step so that each crest and trough cancel out to darkness (**13 C**). This could be taken as a simple "in-line superposition" of total crest-trough cancellation.

Multiple wave trains traveling *in the same direction* maintain the integrity of their specific amplitudes (a) and frequencies (f). However, these can be S*uperpositioned* by algebraicly blending each one into a single periodic pattern called a "vector wave".[14] **Figure 14** shows a "vector" composite C of the two waves A and B. Such vectoring sim-

Figure 13 "In line" Wave interference

plifies calculations, but leaves out or "smears" much of the specifically detailed information of each independent wave. However, reversing the algebraic process might "re-constitute" each original individual wave-train. Waves of different frequencies and/or amplitudes can be superpositioned like this *only if traveling in-line;* but "quantum superpositioning" can be applied to any two or more states, or to other kinds of systems. A

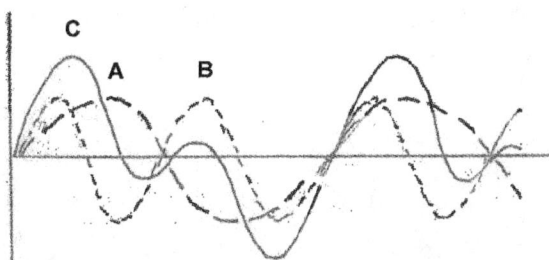

Figure 14 Superpositioning (C) of two waves A, B

somewhat different kind of "superpositioning" can be applied to wave-trains or wave-forms intersecting at different angles as *interferences* other than "in-line", such as those referenced in this work.

Although not usually represented here, keep in mind that all material surfaces, barriers and objects are also made up of patterns of closed and standing spheric waveforms of varying complexities as in **Figure 15.**[15] This would apply to the above sections on *reflection, refraction, and especially diffraction,* which are interferences of wave-trains with complex spheric matter-wave structures. The photo in **Figure 16** is described as "diffractions of high frequency EM photon particles from the atomic-molecular lattice structures of the Platinum target object"; but *any* beamed EM waves are basically *interfering* with the *wave complex of* the *target object,* resulting in secondary ("diffracted") moiré patterns produced from the wave interferences as shown in the photo. The observed phenomenon in the photo is the same, but the *interpretative descriptions* have been changed in keeping with *Energy as fundamental to matter, and as wave is primary over* "particle" (See Figure 17 A,B,C,D below). The caption for Figure 16 read that "The white dots are atoms" which is obviously not true. They are just interference moirés as can be seen in the enlarged inset.[16] The "particle" interpretation leads to contradiction and paradox, while the wave interpretation clarifies and resolves many of those, and is more coherent and *fundamentally non-duallstic*[17]. In this presentation, all "particles" are seen as *standing spheric wave-forms*, pulsing from empty Energy-Wave centers as in preceding figures 7 and 8, perhaps from something like the hypothetical "zero-point vacuum-energy state" of the EWF where the inward pulsing waves emerge from—and the waves pulsing outward are absorbed into—The Energy Wave Field as "the underlying reality of the space-time Universe". It might also be that the outward pulsing waves are directly connected to the inward pulsing waves of its "opposite twin", somewhat like the electric "+" and "—" poles, and the magnetic flux poles would be connected in 3-D as directional flows of their respective energy fields. These oppositely pulsing pairs appear to be basic to Quantum theory (See also Figure 35).

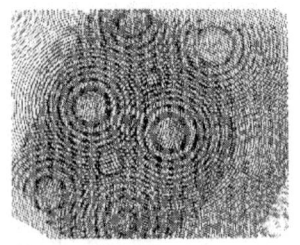 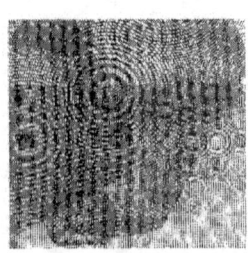 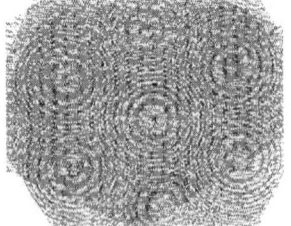

Figure 15 A, B and C matter as complex spheric wave interferences

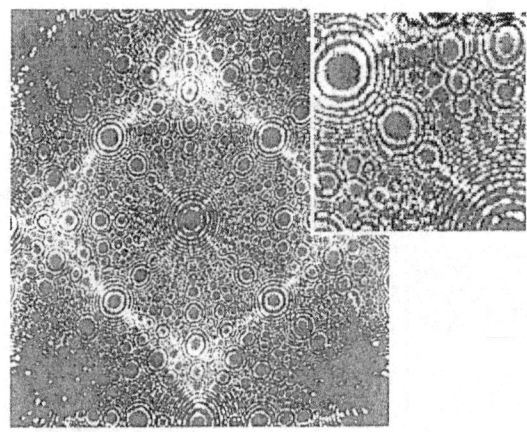

Figure 16 Platinum at ten million X magnification

The most interesting interferences occur when two or more wave-sets interfere at less than 45 degree and more than 0 degree intersect angles. These produce *moiré patterns* as secondary traveling energy wave-trains. **In Figures 17 A and B,** the small arrows indicate the direction of motion of one of the *interfering wave-sets*, and another arrow shows the *direction of travel of the secondary* wave-train. Also produced are the standing concentric or spheric "matter" wave-forms that constitute our material world. In **Figures 17 C and D**, labels indicate the paired *inward and outward pulsings* of the secondary matter-waves relative to the direction of motion of the interfering wave. What is most interesting and revealing are these secondary oppositely pulsing pairs of spheric-matter-waves that result from the interactive interferences of *a radiant wave train with a spheric matter-wave.* The opposite pulsings are dependent upon the angular motion of the radiant Energy waves relative to the initial 2-D spheric wave. This effect corresponds to any quantum "particles" of "opposite charges" and to "matter / anti-matter" and "quark / anti-quark" entities.[18] Like the polar charges of an electric force field, the paired "in and out" pulsings might be connected in a 3-Dimensional field-circuit. All of the figures in 17A and B will be pertinent when later examining the Compton effect.

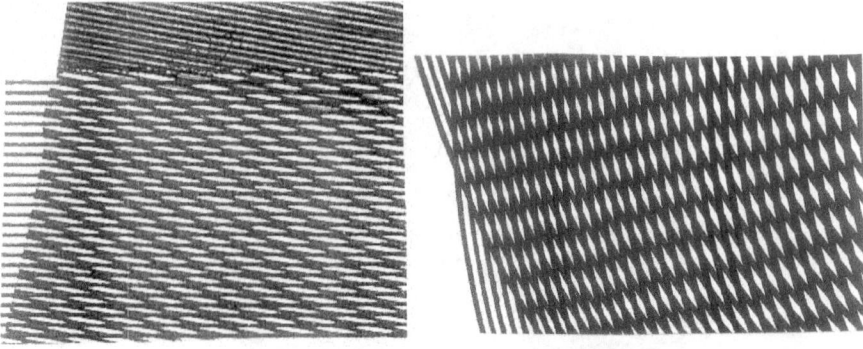

Figure 17A Rectilinear wave interference   **Figure 17B Curvilinear wave interference**

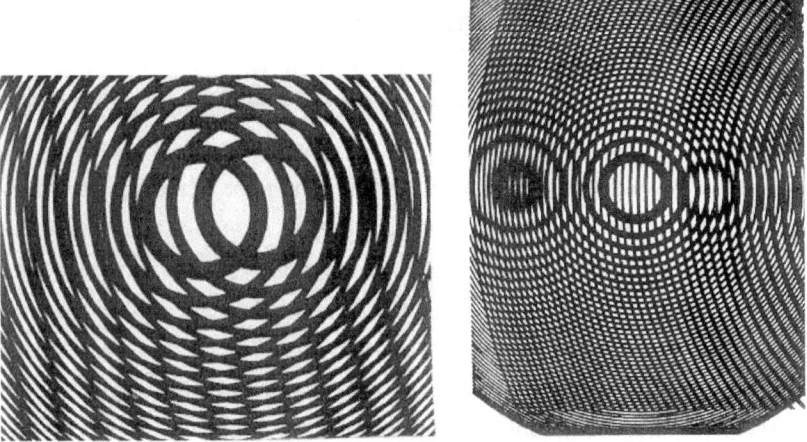

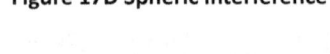

Figure 17C Two spheric waves intersecting.   **Figure 17D Spheric interference**

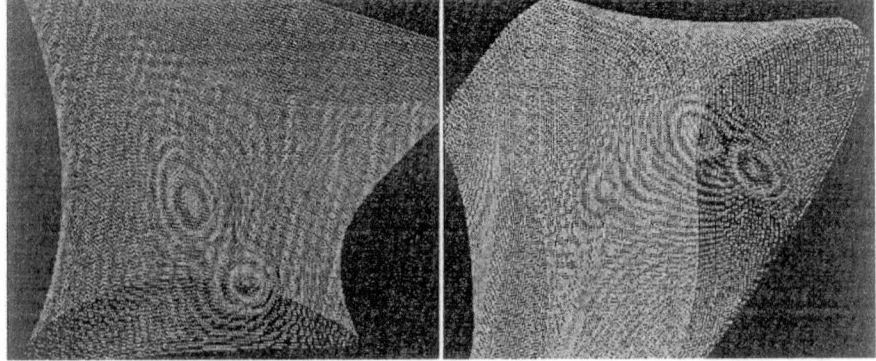

Figure 18A and B Ovoid wave-forms from self-interfering rectilinear wave-trains

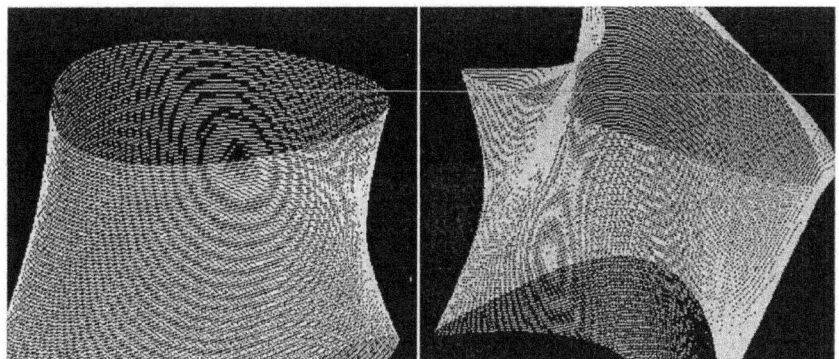

Figure 18C and D two more ovoid wave-forms from self-interfering rectilinear waves

I had earlier assumed from transparent overlays that rectilinear or curvilinear wave-train interferences could only produce secondary *wave-trains* and not *concentric or spheric wave-forms*, but I have since discovered on my computer's "screen-saver" (common to most Windows programs) that recti-or curvlinear waves *do* produce 2-D spheric wave projections as shown in **Figures 18 A, B, C and D.** Except for being a single Möbius-like band instead of a closed spheric-wave form, these are nevertheless good examples of the convoluted *twistings and warpings of a single wave-form folding into self-interference.*[19] (to see in motion, go to "*my computer*" and "*control panel*", then to "*appearance and themes*". In "*screen-saver*", select "*Beziers*; *length—loop, "2" (Width— "loop repeat 85,"* and on *slowest possible* and observe.).

THE QUANTUM

The idea of the "quantum" was first posited by *Max Planck*, who believed that quanta existed only when Energy interacted with matter, but that *Energy itself was not quantized. Louis deBoglie* pointed out that energy and matter *both* appeared to have wave and particle characteristics. *Erwin Schrödinger* was the "father" of *Quantum Mechanics*, (QM) and believed that waves were real, but *Max Born* believed that waves were only *waves of probability*, and from that came the idea that the "wave function" consisted of probabilities of all possible outcomes. When an actual measurement or observation was made, that probability wave function "collapsed" to what was actually observed or measured. All of these varying interpretations of just exactly *what* was observed resulted in several "schools" of quantum theory as to exactly *when* the wave-function "collapsed".[20]

One that was favored by *Max Born*, was that *knowledge of the collapse occurred only in the mind*, and that knowledge *constituted any knowable "external world"*—a kind of *"all-in-the-mind"* theory that perhaps harked back to *Bishop George Berkeley* (17th-18th centuries) who taught that the external world has no existence independent of concepts in our minds as "sparks of the mind of God".

A somewhat similar idea was that held by *Eugene Wigner*. The wave-function collapsed when a *conscious observer observed the result* of an experiment or when making a measure. The main difference from the first "school" above is that the *conscious observer observes* a real world that exists *external to and independent of conscious observer*.

A third and the most "popular" idea favored by *Neils Bohr* is known as the "Copenhagen school". The idea is that there are *two distinct worlds*—the *micro-world of the quantum, and the macro-world of Newtonian physics and Einstein's relativity*, and that knowledge of the micro world consisted *only* of "clicks" on a

Newtonian macro-world Geiger-counter or the numbers indicated on a measuring instrument. The "collapse of the wave function" occurred at the transition between an unseen quantum micro-world and the instruments of our familiar Newtonian macro-world. *All we know are basically numbers and relationships of numbers in equations.* (See "The Real Number System" on page)

A fourth hardly merits mentioning, the "many-worlds*" theory*. Since the wave-function consists of a superposition of possible outcomes in a decisive observation, the selected outcome remains as it were in *this* world, while *all the alternative outcomes* (or decisions or choices) *still exist, albeit in another world or in other worlds*. The absurdity is that every time any alternatives are *not* observed or selected, there are new worlds formed where those alternatives exist, and we as "observer-choosers" exist with them as "alternate selves". This view was suggested by *Hugh Everett III* in the 1950's. Of course, it is completely untestable.

A fifth idea as put forth in this paper is the *Energy Wave-field* hypothesis where *Energy Waves are real but unmanifest until self-interferences manifest in secondary effects* as radiation wave trains and as matter wave forms that are actually *standing spheric waves* commonly mistaken as "particles". This mis-identification seems to have originated with *Einstein* during his photo-electric experiment which involved bombarding a metal plate with EM radiation. *Only a critical minimum energy frequency* would "dislodge" the plate's valence electrons from their orbit shells of the plate's atoms. The "liberated" negatively charged electrons could then be measured as an electric current. (See Figure 22 below under "Photo-electric effect")

"Particles" were not a part of Planck's original "quantum". "Particles of energy" were entirely Einstein's idea, and almost none of his collagues agreed with him. In his original paper, Einstein's 1905 equations initially contained a *separate but necessary part for the dichotomy of wave and particle*, stating that light and all energy behaved "as though" it consisted of quanta.[21] But having already been an advocate of atoms and Newton's "corpuscular" theory of light, he believed that the wave-particle duality would vanish if energy—like matter—could be treated as discrete particles.[22] Knowing of Planck's hypothetical "quantum as packets of energy", yet *contrary* to Planck's belief that quanta appeared *only* when energy interacted with matter, Einstein asserted that the incident Light rays *themselves* came in discrete "quantum particles of energy". In 1916, *Leonard T Trolard*[23] first dubbed these hypothetical particles of light as "photons" (from the Greek *phos* = "light"). For almost 20 years, most physicists did not accept this idea, but by 1922, *the Compton experiment* described below was believed to be the final "nail-in-coffin" evidence against wave theory, and Einstein's "quantum particle" of energy, i.e., the "photon" was finally accepted by most of his colleagues.[24]

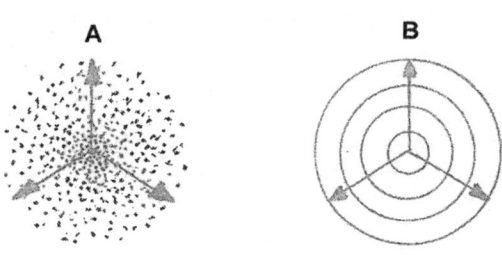

Figure 19 Radiation expansion of A Einstein's particles; B Wave expansion

Einstein argued that *"According to the assumption considered here* (i.e. his quantum-particle hypothesis), *in the propagation of a light ray emitted from a point source, the energy is not distributed continuously over ever-increasing volumes of space, but consists of a finite number of points of energy quanta localized at points of space that move without dividing, and can be absorbed and generated by matter only as complete units*—which is a profound formal *difference* [from waves] ."[25]

In other words, waves that are "continuous" will "spread out" over an increasing volume. But how can that be any different from the thinning distribution of a "finite-number of points localized in space" that spread out over "ever-increasing volumes of space" as the radius of propagation increases? (**Figure 19 A and B**) Being "generated and absorbed only as complete units" is the same as an indivisible "wave packet of a fixed quantity of Energy", which is *not* equivalent to a "particle"! "Photons as *localized points in space* is pure conjecture, and photons that can be *absorbed and generated by matter* is just what Planck had *already* maintained as *'packets of energy"*, and *not* as "particles". The thinning distribution of "points spread out over an increasing volume" is no different than that of waves. Looking at *Christian Huygen's "principle" of 1690*, he posited that as waves propagate, each *point* on a wave-front acts as a new source of disturbance. That same principle would hold for Einstein's *"discrete points in space"*, but would not be the same as "particles of energy". Huygen's "points of disturbance" are his geometric theory for how waves continue to propagate from one central source of disturbance when *there is no physical medium*. (**Figure 20**). Einstein's *profound formal difference*, argument just does not hold. But Huygen's principle might apply to all waves that do not propagate through a material medium.

It is reported that Einstein also maintained that the energy in a classical wave is dependent on *its <u>intensity (i)</u> and <u>not on its frequency</u>* (f).[26] This statement is in

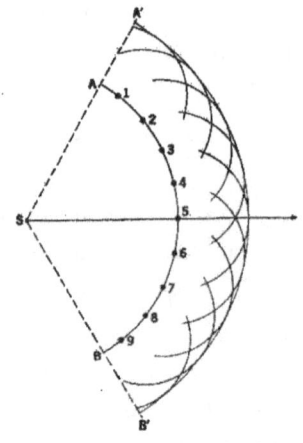

Figure 20 Huygen's principle for EM and Energy Wave propagation

abject contradiction to what was then a known fact. In the EM spectrum, the higher-faster frequency vibrations have *more* energy than the lower and slower frequencies, and *that energy is independent of any intensities.* Energy that is manifest as *intensity* (*i = amplitude squared)* has a different effect from the energy manifest as frequency. In the photo-electric effect, the intensity determines *only how many electrons* might be emitted, leading to Max Born's "rule" that *intensity gives a* "cloud of probability" as to *where* an electron would likely be found. However, the energy required for the photo-electric effect to occur at all *is the critical minimum frequency,* and Einstein had to have known that from his experiments. In the photo-electric experiment, it was the *energy of frequency* that determined a "yes / no", or a "go / no-go" of *electron ejection,* as well as the *Ek velocity energy* of the ejected electrons, and this is incontrovertible and well established. (An aside: the Hydrogen atom's electron as a "*standing wave*" is shown with many theoretical *wave-function configurations* as in **Figure 21**.[27]

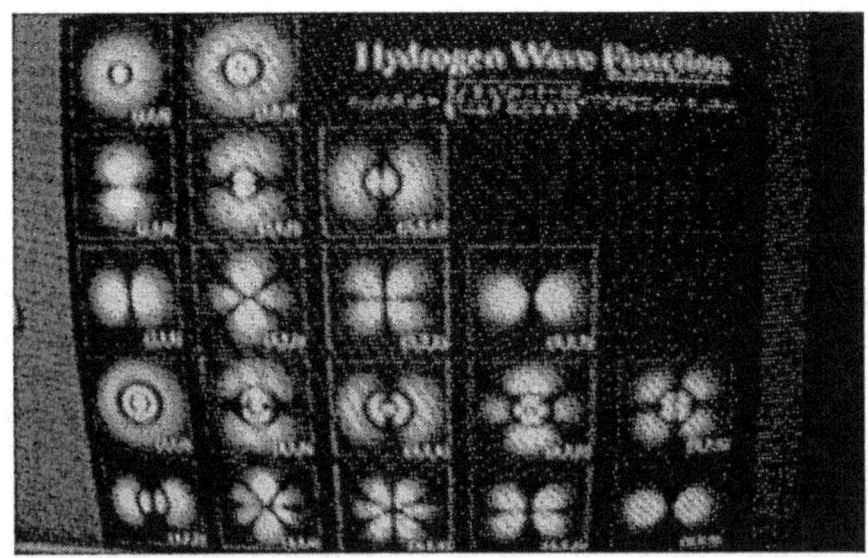

Figure 21 Hydrogen Wave Configurations

Photo-electric Effect

In the photo-electric effect, the "critical frequency" is relative to the stability of a particular metal's valence electrons. Einstein maintained that the *"wave-like phenomena of Light appeared over time, whereas the instantaneous emission and absorption of Light [as quantum-particles] was quite different, and that this made clear why wave theory failed to explain the photo-electric effect"*. A "photon-particle's" *intensity* could only be the *radius of its oscillations*, and the "photon's" *frequency would* be the number *of oscillations per second at which a photon vibrates.* Since frequency is time, *how is that different from the frequency of n—waves per second?* At or near the speed of light, a sufficient number of oscillations of a single particle would take the same interval of time as a critical number of wave crests to strike and effect the electrons of the metal plate. *The critical frequency as time would be the same.* Einstein's reasoning doesn't make any sense, and seems more like subterfuge to gain acceptance for his "hypothetical particle" to *replace* the wave theory of light and energy. Einstein called the quantum a "particle of energy" vibrating at a frequency of n-oscillations per second, but in his arguments, *there is a not-so-subtle shift from a quantum as a packet of energy to a quantum as a particle*, The "quantum" as "packet of energy" was *never* thought of as a "particle" before Einstein's assertion. The new course was now set that Einstein later repudiated.

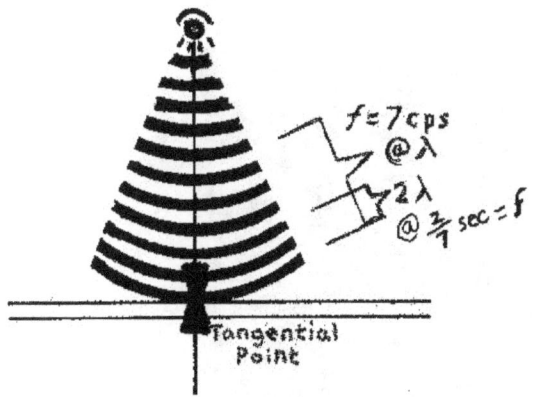

From most artificial light sources, wave-trains are usually *curvilinear*; straight rectilinear waves might radiate from fluorescent tubes, but most light radiating from spheric emission sources, or from lens and reflectors that focus light appear to radiate as curvilinear waves-trains. The curved "+" crests initially strike a flat metal plate at a tangent to the surface, and each successive wave strikes staccato-like. I believe the later successes of photon-particle predictions is primarily due to the relative ease of calculations as compared to wave calculations, and in **Figure 22,** the *tangential point of incidence* of a curvilinear wave-train can be and was interpreted as a "particle" or a succession of particles. However, even with successful predictions, the

Figure 22 First two incident waves at wavelength "instantly" gives the full frequency

particle idea seems to involve contradictions. As shown in Figure 22, the wave theory *can quite adequately* explain the photo-electric effect. Every one of the emitted curvilinear "+" wave crests initially strikes the plate at a tangential point in a *stacatto effect*, and *the time unit for both wave frequency and particle oscillation frequency is the same one second,* so Einstein's "instantaneous" time factor is a moot point. Einstein *insisted* that *all* energy itself came *only* as "quantum point-particles". Since wave crests are "+" signed, at velocity (v), the distance (L) between any two adjacent periodic waves striking the plate already determines the over-all frequency, so it might take only the wave-length spacing of any first two incident waves to "instantaneously" give the entire critical energy-frequency of the EM ra-diation for the photo-electric effect to occur. But even if not, *at or near the speed of light, how instantaneous can you get,* especially given *Planck time* as 5.39124 x 10^{-44} seconds? In any case, *since light-speed is the same for particle or wave*, Ein-stein's "instantaneous" wave-particle time-interval argument seems *specious*. Even when Louis de Broglie showed that particulate matter also behaved as waves, Ein-stein still insisted on the "particle-ization" of light and of all energy. (In Figure 22, the frequency numbers of waves are *arbitrary* only to illustrate the wave packets and their tangential points).[28]

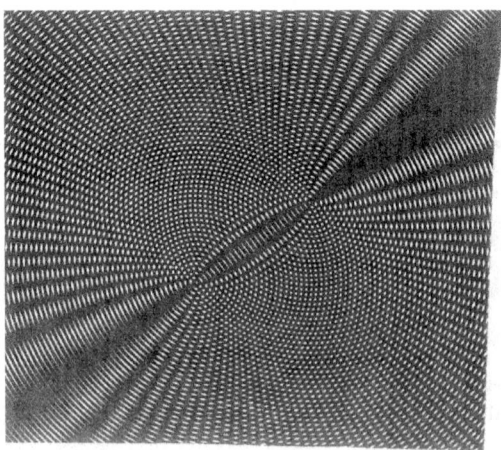

Figure 23 "Vibrating String" moire from two off-center centric waves

All vibrations or oscillations at some fundamental frequency have *harmonic multiples* of the fundamental. If anything, a "quantum" is more like a *fundamental vibration with multiple harmonic overtone vibrations,* each a universal indivisible ratio, but never as a "point-particle." "String theory" seems to be based on such an idea. But as Einstein transformed the quanta from indivisible "packets of energy" into zero-dimensional "point-particles", the string theorists have transformed those "quantum point-particles" into tiny one-dimensional vibrating "strings". These invisible strings require not only the three dimensions of space and one of time, but as many as an additional seven, fourteen, sixteen or twenty-two dimensions, depending on who you are reading that are all curled up and invisible. **Figure 23** looks very much like a vibrating string, but is in fact an *interference configuration* of two concentric wave-forms with their offset centers acting "as though" they were distant end-points.[29] Many other interferences and their moiré patterns suggest EM "signal beams" (**Figure 24**), lines of *electric force-fields or their equipotentials*, or lines of *magnetic flux.* (not shown here). Many invisible phenomena depicted graphically in textbooks actually do represent wave interference patterns. Though the all-pervasive *Energy Wave Field* is itself unmanifest and invisible, the self-acting interferences of Energy waves *directly manifest* as natural phenomena.

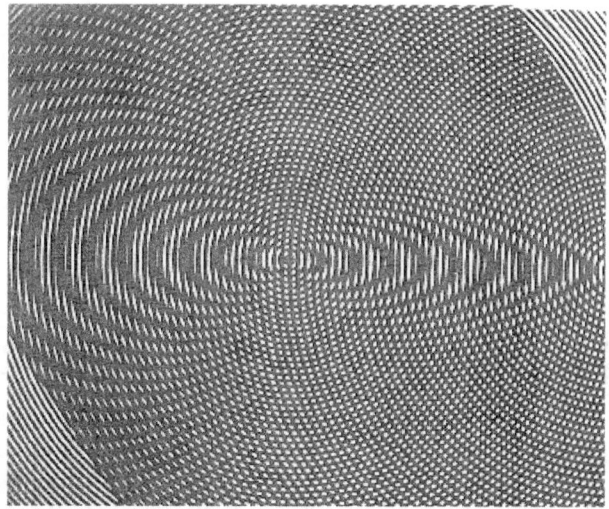

Figure 24 "Signal Beam" moiré from two off-centric waves

The Compton Effect

For almost twenty years, many of Einstein's colleagues tended to resist his idea of energy itself being "quantum point-particles",[30] until the American physicist *Arthur Holly Compton* in 1922 was testing the effects of X-ray beams aimed at and striking various elements such as carbon. When the experiment was performed, X-rays that were "deflected" at various angles of "scatter" were measured for their frequency against that of the initial X-ray emission. Based on classical Newtonian physics, the then current wave theory apparently predicted the frequencies to be the same; but if the deflected X-rays had a *lower* frequency with a corresponding *increase* of wavelength compared to the initial incident rays, it would mean that the X-rays could *not* be waves, but were *deflected photon-particles* instead . When this was in fact observed, it was taken as "irrefutable evidence" in support of Einstein's "quantum particles of energy,[31] with wave theory relegated as a subordinate theory. But the use of the terms "scatter" and "deflection" *already indicate a bias* that x-rays would act as "particles bounced off of matter-particles" and not as *energy waves interfering with matter waves*. However, the "irrefutable evidence" in **Figures 25 A and B** clearly shows that *when two wave-sets are in interference, the resulting secondary wave set is unmistakably of a longer wavelength and lower frequency* than the initial interfering wave-sets, which is just what Compton said could *not* be waves. Contrary to repeated assumptions and assertions, as described by the wave actions in figure 22 and interferences in figure 25, the wave theory does *quite easily explain both the photo-electric and the Compton effects*. Note that in 17 A and B, the resultant second-order rectilinear wave-trains labeled "c" travel at a longer wavelength and in a distinctly different direction from either initial wave-set as shown by arrows; that secondary wave angle is determined by *the angle of incident interference of the two interacting wave-sets.*

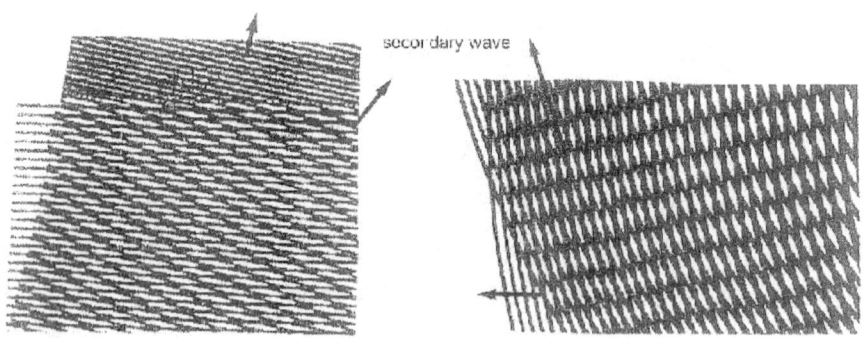

secondary wave

Figure 25 A and B Compton effect with longer secondary wavelength interferences (arrows indicate direction of overlay motion and secondary wave motion)

In traditional classical physics, an electron "particle" can have a "trajectory" as successive static positions during a specific interval of time as a *simultaneous measurement* of position and velocity ("Velocity" is a vector of speed and direction). In Quantum mechanics, however, an electron can theoretically *have no trajectory*, in part because of *Werner Heisenberg's* "Uncertainty Principle" about the inability to simultaneously measure any two complementary pairs of observables, such as position and momentum, and partly because for Max Born, an electron is only a "wave of probability" with no specific position. However, in *Particle Physics*, "particles" in high speed accelerators are sent smashing into each other, leaving very definite trajectories *on film* of the aftermath of these "collisions" (See **Figure 26).** Also, in various kinds of "cloud chambers", "particles" leave a kind of "vapor-trail" trajectory consisting of condensate droplets. These are unmistakable *observations*, of tracks and trails left by *something*, most likely *spheric wave-forms*.

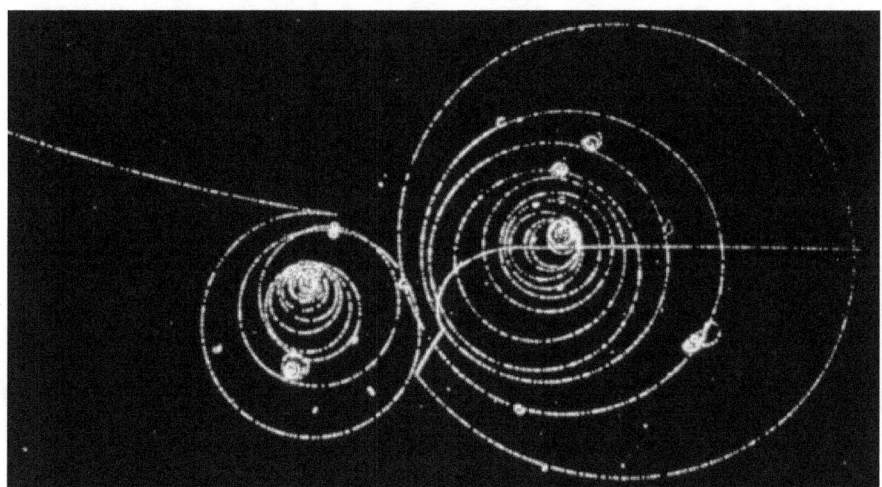

Figure 26 Spheric wave (ie. Quantum particle) trajectories on film

What happened to Einstein's "photon-as-particle-of energy" is just one example of the ad hoc adjustments applied to uphold an entrenched "standard theory" that has become problematic. $E = mc^2$ means that *all* Energy is *mass* and is, has or exerts *gravitation*. But anything traveling at the speed of Light (c) apparently cannot have mass, which is probably why the "photon" was declared *massless*. Yet the photon still has momentum $(p)^{32}$, which in classical physics is *mass* x velocity. *Louis de Broglie* used *Planck's constant (h)* divided by *wavelength* L (for the Greek "*Lambda*") to be the momentum of the "massless" photon. As all Energy—including EM (Light) energy—is mass, a massless photon could then no longer be "energy itself", and had morphed from a "particle of energy" to a massless "carrier of energy"; later, Einstein's "photon" was suggested to be "*guided by a pilot wave*"[33] bringing the whole sequence of ad-hoc adjustments full circle. But here is the real point: *Einstein* and *Neils Bohr* had a long-running debate concerning the relationship between "reality", "probability", and the "observer', and at the 1930 *6th Solvay Conference*, Einstein presented to Bohr a "mind experiment" that initially stymied Bohr and his position. The "thought experiment" involved the weighing of a contraption before and after a single "photon" had been emitted from an aperture. The point is not the debate, but that *as late as 1930, both Einstein and Bohr accepted that the photon had weight and therefore had mass.*[34] But if Einstein's E = mc^2 is correct, then *ALL Energy is mass, and Light, being EM energy, must also have or be mass,* which means that *"photons as Light" cannot not have or be mass*: I believe the problem is primarily due to Einstein's idea of energy as *photon-particles.* There is also an irony here in that Einstein, who was the "protagonist" using what appear to me to be specious arguments for "quantizing" all energy into particles, later decried the way quantum mechanics was developing. As the photon was deemed massless, it could no longer *be* energy and had to be a mere *carrier* of energy, and it must have been these kinds of modifications as quantum physics was developing that troubled Einstein, a questionable "balancing" act" between requirements of physics laws and those of the standard "quantum particle" model. Yet it was Einstein himself who had initially *insisted* that energy itself was "quantized" with the quanta as a "particle" instead of an indivisible "packet of energy", which later led to contradictions and subsequent ad hoc "adjustments".

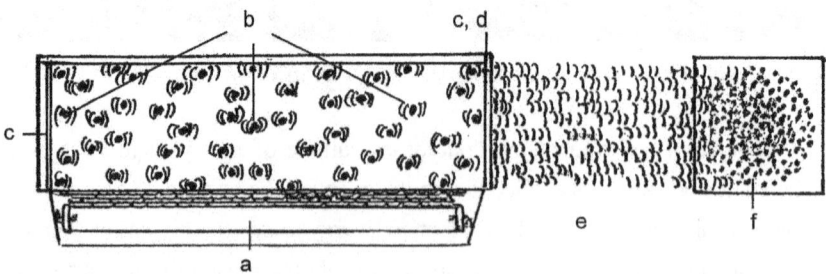

Figure 27 Pulsing laser beam: pulsing "stimulation lamp" (a) *excites electrons* (b) to emit radiation pulses *amplified by reflection* back and forth between mirrors (c) to burst from front mirror (d) *in pulses* (e) as "*scintillating flashes*" (f) on film

When a laser light is shone onto a smooth white paper, even with the naked eye, the spot on the paper appears to consist of "scintillating points". Were these actual "particles of light"? Under magnification, it was evident that these tiny "particle-like" scintillating points did not change positions; but flickered in place, and it suddenly occurred to me that *LASER* stands for *"Light Amplification by Stimulated Emission* of Radiation". Ordinary "incoherent" white light consists of a visible spectrum of frequencies perceived as *red, yellow, green, blue and violet* when shone through a prism. Laser light, however, is "coherent" and consists of a single frequency, usually from a rod-like crystal that is mirrored at both ends, the "back end" more heavily than the "front end". *The crystal must be stimulated by an external, usually rapidly flashing light source*[35] (**Figure 27-a**); the "excited" electrons of the crystal's atoms absorb the light's energy pulses, then emit that energy at the crystal's frequency (**b**); the mirrored ends reflect all those electrons' emissions back-and-forth repeatedly, *amplifying* the laser light energy until sufficient momentum enables a beam (e), to burst forth from the more lightly mirrored front end (**c, and d**).

The atoms in a "solid", especially a crystallized solid, are "locked" into a lattice-like geometric structure, and can only perhaps oscillate slightly in place, but the atom's *valence*-electrons absorb and emit energy in "pulsing flashes" as the external light stimulates in pulsing flashes. That is why the scintillating points (f) all remain in the *same place* on the paper, and why they appear to "sparkle and scintillate" as pulsing flashes. These *appear as though* they are "photonic particles of light", but they are not; between the relationships of the external stimulating light, the intermittent absorption and emission by the electron waves and the crystals mirrored ends, the observed "scintillating points" *could be* "quanta" as *"packets of energy"*; but *not* as actual "photon-particles". A *packet of energy* is a pulse or series of pulses of waves as depicted in the projected beam (e). *A "wave packet" would be the frequency, so a "quantum" should be the ratio between the frequencies of the incident, the absorbed, and the emitted radiations. There are no* "particles of energy"; *nor, for that matter, are there* "particles of matter."

According to quantum theory, both energy and matter behave sometimes like "particles" and sometimes like waves"; in interactions between matter (electrons) and electro-magnetic energy ("photons") a "wave-function" will "collapse" into a particle. But in 2007, a team at *Lawrence Berkeley* headed by *Greg Engel and Graham Fleming* made a break-through in *photosynthesis*. "Long-term quantum coherence" is when sunlight's broad spectrum of wave frequencies allows the molecular cells in plants to "select" the appropriate wave frequencies needed for photosynthesis. *Coherence* means *that the wave is sustained and does not "collapse" to a "particle"*[36] The matching of the molecular "need" with particular frequencies of sunlight is called *harmonic resonance*, and I believe it is that interaction between energy waves and matter waves that Planck referred to as "quantum packets of energy" and not as the "energy particles" that Einstein so insistently proclaimed. In this paper it is maintained that *there is no collapse of a wave-function into a particle, and that particles of both energy and matter do not exist;* instead, they are pulsing spheric-wave forms that interfere with other spheric matter-waves, and/or with electro-magnetic waves that effect secondary wave configurations as shown in figure 25 A and B, and in figures 35 A and B. The "quantum" as a "wave packet" is not at all in question; only the quantum as a "particle" is being questioned and rejected. For nearly 20 years, almost all of Einstein's colleagues rejected Einstein's idea until the Compton experiment was thought to uphold Einstein's "light particle". But this was already proven wrong.

The Field

The concept of "field" is the measure of some physical attribute that is spread out over every point in space-time. *Paul Dirac's Quantum Field Theory (QFT)* allowed instantaneous *"action-at-a-distance"* which was incompatible with *classical Newtonian physics* as well as with Einstein's *Special and General Relativity theories.* But in any field theory, everything *in* the field is inextricably inter-connected, regardless of any distances involved. However, *within* QFT, there are more than one single field; there is the *Q-Electro-Dynamic field* (QED) of Electro-Magnetic energy; the *Q-Chromo-Dynamic field* (QCD) is for the "color" aspect of "quark" entities; and *Q*-Flavor-Dynamic field (QFD) is for "quark flavors"—which is just a name for another set of "quark attributes". Another field has apparently been added, the *"Higg's Boson field'* for the once-proclaimed "God-particle". I've read of a *spinor field* of nuclear spins, and a *twistor field* (apparently having to do primarily with the mathematical changing of sign values, although I can find next to nothing more about either one); like particles, fields seem to proliferate, adding to the existent confusion regarding Quantum Theory. It is admitted even by quantum physicists that—despite its apparent predictive successes—*no-one* really *understands* Quantum Theory. But The Energy Wave Field is a single field, a totality of unmanifest Energy, so that there is no "empty" space, and it is that singular Energy Field in which self-acting Energy Waves, via *Möbius-like twist* and bend,[37] are able to self-interfere, and it is these wave interferences that are the first interference moiré configurations that also interfere repeatedly to form the entire phenomenal Universe. In the EW System, there are regional or local "fields of interest" regarding particular interference configurations as local "systems", but these are not separate from the over-all EW Field; they are but focii of *manifest interferences as phenomena*. Quanta might be real as *ratios or harmonics*; but "quanta as *"particles"* are not real, being instead *spheric wave-forms* of the EWF as illustrated in the "wave primer" section and in Figure 35. Within the quantum program is the idea that—along with energy—*space itself* is also "quantized". Unless I'm wrong, this means that every possible "point" in space acts as an "oscillator". Even if that point were an oscillating "particle", how could that be distinuished from "gravity"? Would oscillation be an inherent characteristic? Or would some outside agent— *energy* perhaps—*make* it oscillate? If you have Energy, why is an imposed metric of "oscillating points" necessary at all? What possible purpose would they have, except to use as positional coordinates? Then why oscillate? And does that reflect "reality"? With the present attempt to quantize gravity to complete a *Grand Unified Theory* (GUT), if energy and space-time are already "quantized", and if gravity is the geometry or architecture of space-time, would not gravity already be quantized? Or is the whole endeavor of "quantizing" everything merely a labor of Sisyphus? Might thought itself be "quantized" into bits and pieces? It seems already to be. The main characteristic of any "field" is that all "events" are inextricably interconnected. Consciousness and rational thought is just such a field. "Entanglement" is a primary feature of the *field concept*, not only of quantum theory.

The Real Number System, Arithmetic Rules and Procedures[38]

Since most if not all of physics is based on mathematics, it needs to be kept in mind that *mathematics has its own rules for each subsystem of the Real Number System*, and although these rules are *independent* of nature, the system and its rules are "borrowed" by physicists to express relationships *in* nature. Their faith in numbers is inherited from the ancient followers of *Pythagoras*, who believed that nature and the universe were rational and written in the language of number and proportion. But as the number system and arithmetic rules are *self-referential*, they may have no actual relationship to nature at all; when measuring quantity, magnitude and proportion in nature with numbers, since the measure is number, *the results we get are always numbers from that same mathematic system of rules*. The question is "Are we calculating nature, or just playing with numbers?" OR, is there any difference?

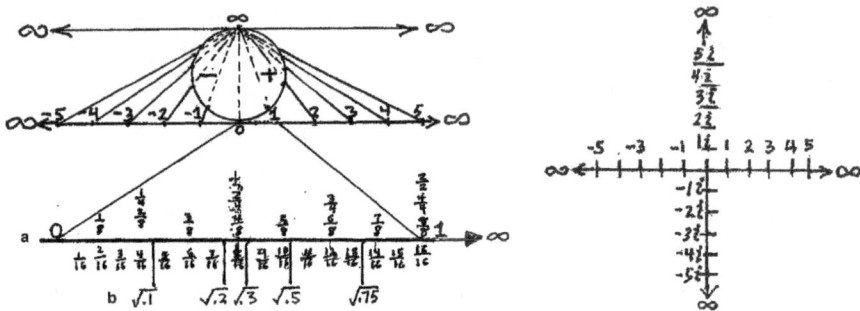

Figure 28 A The Real Numbers with fractions (a) and unending decimals (b)
Figure 28 B Imaginary Numbers (vertical axis) with Real Numbers on horizontal axis

The *Natural or Cardinal Numbers* are all "positive" (1,2,3,4 etc.). Adding the negative numbers to allow for subtraction makes up the system of *Integers*. These are often shown as arranged on a horizontal line that extends out to "infinity" at both left and right ends. In the middle is "zero" as the beginning of each sequence, the negative to the left, and the positive to the right. The representation of this is shown in **Figure 28 A**. In between any two integers are the *fractions* (a) which make up the *Rational Number System* that includes periodic decimal sequences written as fractions (e.g., .333... as ⅓). The *Irrational Numbers* consist of non-periodic, infinitely random decimals that cannot be written as fractions (e.g., the square root of 2). These are represented as points that form an *infinitely dense* arrangement of positions. The addition of the unending decimals that cannot be written as fractions completes the *Real Number System*. These are the *Irrational Numbers* (b) that consist of "periodic" decimals that endlessly keep repeating a sequence ad infinitum, and other "non-periodic" decimals that seem to be infinitely random. The 'Irrationals" are packed even "more infinitely dense" between all the "rational fractions" (if it is possible to say "more infinitely"). The first Irrational Number to come to the attention of the Pythagoreans was the *square-root of two*, and it shook their faith in the natural order of things. This completes the *Real Number System*, but the definition and arithmetic rules applied to physics theories of nature sometimes make for what appear to be strange puzzles. For example, according to one of the rules of multiplication, a "+" number and its "—" twin when multiplied together yield a "—" product regardless of the quantities. In the wave theory of light, the wave diagram has waves that peak above a median line of propagation as "+", and that descend an equal amount below the median line as "—" (See Figure 1). Each represents the *amplitude* (a) of light, and light's *intensity* "would be" the product of the "+"crest and the "—" trough, but the product would be a minus number: therefor instead of *multiplying* the "+" and "—" values the signs are ignored, when *squaring* a number, so the intensity becomes $i = a^2$. But the median line still represents "0" (if added) or minus (multiplied) between "+" crest and "—" trough, as well as *the speed* of light's propagation! This is just one example of the kinds of contradictions that occur when applying numbers and mathematical rules to nature. Another example is a movement from "A" to "B" over distance "D" yields a "+D" number. But *moving back* from "B" to "A" negates that number: +D = A + B one way, and —D = B — A in the opposite direction back again to the beginning point with the *total distance traveled = 0!* That's the result of the rule of signs applied to direction of travel from the order of integers. Another example of "the rule of signs" is that the *product* of "—" times "—" is a "+", and that *cannot be represented* graphically. Does applying the "rule of signs" to nature and her processes tell us anything about *nature*, or just about the *rules of mathematics*? How much of our "knowledge" of nature is *determined* by the arithmetic rules themselves? The supposition that nature is "written in numbers" is because numbers are used to measure nature, and since the rules are self-referential, when *we find numerical relationships that seem "magical" or "astonishing", or that numbers somehow always "come out in the end", are we seeing nature or our own number-system-as-measure?* Even in the unscientific "occult" numerological methodology, "astonishing magical relationships" seem to occur that are actually between the numbers themselves, and it seems to be the same thing in the scientific use of mathematics. When the "numbers all "seem to come out", *it is because the rules of number relationships always beget numbers!* The number system is a more or less *logical symbolic construction that*, instead of reflecting relationships in nature, may only reflect self-referrent relationships of the mathematical rules as "logically constucted".

A note on "Infinity," of which there are two kinds: one is the un-endingness that can never be complete and is always "unfinished"; the other kind is the *truly Infinite* that is always a *complete and whole Totality*, one that holds *all* within itself.

(The word "*whole*" is derived from "health" and "heal", and related to the word "*holy*".) The unending kind of "infinity" is a *specious* infinity; it is the "infinity" of the positive and negative numbers that continue outward from either side of "0" and that never reach completion; the fractions and the "irrational numbers" and the *infinite number of points* that can be placed between any two integers are of the same kind of specious "infinity" as unending incompleteness. Infinity: totality vs. the incomplete.[39]

In 1931, the mathematician *Kurt Gödel* proved that *every system contains statements that cannot be proven within that system, and that no logical system can prove its own logical consistency.* A statement may be true, but cannot be proven, and what might be proven is not necessarily true. This applies to the system of numbers and all definitions, rules and laws of mathematics.[40] Could it be that these are the very instruments giving rise to "quanta"? Ratios certainly do appear when nature or any system is measured. The mathematician *Georg Cantor* proved that a *line segment* could be mapped onto a square with implications that the dichotomies of "continuous" and "discrete", as well as the "part" and the "whole", and even "dimension" itself are questionable. There are "grammatical" rules for the mathematic language just as for spoken and written language; yet even among theoretic mathematicians, there is disagreement about what mathematics is or means, and its relationship to reality. The number system itself consists of discrete points representing the whole numbers, and finer points representing those infinitesimal fractions and the irrational, infinitely endless decimals-points so infinitely dense as to become indistinguishable from a continuous line. Even with the "continuous" waves, the "+" crests are discrete, and the numbers used as *cycles-per-second* are discrete. The EM waves form a continuous spectrum of *discrete frequencies*, and *even a vibration or oscillation consists of discrete pulses or movements* (See Figure 3). Are any of these discrete pulses considered to be "particles"? The question is "How does a "packet of energy pulses" become a 'particle'?" Our modern conception of a "particle" (atomic and otherwise) is one of a "*mathematical center of force*", and "force" is something that makes something else happen. That "something" and the "something else" is a *metaphysic*. Do the divisions of the "discrete quarks" dissolve into a continuum of emptiness? No physical science can ever be free of some underlying metaphysic, including that of Pythagoras and his followers who believed that all nature was written in numbers. The "calculus" itself chops motion or any changing state into discrete units, somewhat like Zeno's flying arrow; but these "units" can be divided into infinitesimally smaller units that approach but never quite arrive at a smooth unbroken curve. Might "the quantum" after all be merely an *artifact of the calculus*?

On the other hand, since our brain has evolved from the Universe, can our number system as a product of our brain, of thought, be a "Platonic-like idea" (i) that is the "footprint" of the mathematics of the Universe? Both are metaphysical questions, and either one can be a belief or bias, and neither can be answered unequivocally. The supposedly "necessary" use of the *Imaginary Number System* in Quantum Theory' to "adequately describe" *superposition* is questionable to say the least. *The Imaginary Number System* is a theoretic "extension" of the entire Number System, but is independent of and on a completely *separate axis* from the *Real Number System* (See **Figure 28 B**, page). This axis consists entirely of *impossible* numbers—"non-numbers"—based on *the square root of minus one,* numbers that can have no relationship whatsoever to the Universe and the phenomena of Nature. The brain in its power to imagine *anything* seems to have run amok; $4 + 3i$ is neither more nor less than $3 + 4i$, but not equal to $3 + 4i$! Although the "superpositioning" of a number of waves of different amplitudes and frequencies into one complex "vector" wave-system is an "algebraic device" and not a "natural phenomenon", nevertheless using *imaginary numbers* in superpositioning seems like an arbitrary contrivance. "Superpositioning" seems to be nothing more than "averaging" waves, except in "quantized" imaginary minds.

Neils Bohr pointed out that what we "observe" at the "micro-atomic scale" are only clicks on a "macro-scale" geiger-counter, or numbers of some other indicator-scale on physical instruments in the observer's "classical-Newtonian" world. This means that there is *inherent* in Quantum theory a *division into two different worlds*. In every interpretation of where and when the collapse of the wave function occurs, the physical register-indicator, or the mind of the conscious observer, there is an implied "absolute division" between the quantum micro-world and the Newtonian macro-world, and also between an "external reality" and the "internal consciousness" of the observer. But even the "clicks" and the "numbers on a scale" are null without an observer observing them. The only "resolution" possible to this "cognitive dissonance" is to recognize that "*the observer is the observed*" is a truism at all levels.

Over the past century, various quantum physicists have several times stated frankly that "No one really understands Quantum Theory anyway". Penrose wrote that he had to "side basically with Schrödinger himself, and with Einstein and perhaps more surprisingly with Dirac...and to take the view that present-day quantum mechanics is a *provisional* theory"[41] It seems to me that many of the contradictions, the convoluted *ad hoc* adjustments, the puzzlements—most of the "problems" encountered in Q-theory—are primarily due to the *erroneous identification of the quantum packet of energy as a "particle"* and the stampede to be on the "cutting edge of things" has resulted in a kind of "mania to particle-ize" *everything* in the belief that this will lead to the great GUT. To me, this razzle-dazzle of quantum "particle-ization" has, in effect, been one long detour from any meaningful physics and any "*Grand Unification Theory*. I believe it has held back real developements in physics and quantum theory. That there is success in prediction may have more to do with the *mathematical equations* than with the particle theory itself.

COSMOLOGY

THE "STANDARD MODEL"

In the following text, when *Light* is capitalized, it refers to all Electro-Magnetic (EM) radiation waves, and wherever lower case *light* is used, it refers only to that section of the EM spectrum that is visible to the naked human eye. Lower case "universe" is used in general, but capitalized "*Universe*" is used when referring to the totality of manifestations of the Energy Wave Field. I regard the *Universe* by definition to be the "Closed or the Isolated System" of all *systems*. "Energy" refers to the Universal Field of primary Waves, both as unmanifest and as the initial self-interferences, while "energy" (lower case) refers to local energy interferences manifest as specific systems that are subject to entropy and to measure. I usually abbreviate "Big Bang" as 'BB" and "Q" for Quantum or QT for Quantum Theory. As "Energy" is fundamental to all manifest phenomena, I tend to use capital E with lower-case subscripts for all various forms of Energy. For example, *kinetic energy is* E_k, *potential is* E_p, *chemical is* E_{ch}, etc. Why have completely different symbols for forms of a *fundamental Energy* when they can be differentiated by a simple subscript abbreviation that is much less "cryptic"?

As it stands today, the universe is expanding from a hypothetical "BIG BANG" (BB) that happened around

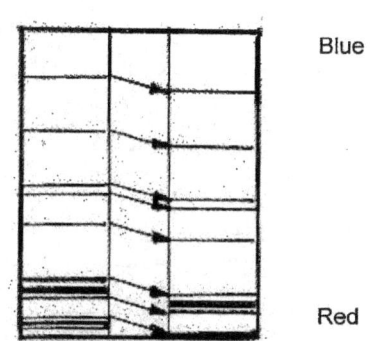

Figure 29 Fraunhofer Lines shifted down (left to right) as red-shifted

Blue

Red

13.7 billion years ago, and is expanding faster the farther out into deep space one looks. From that initial BB, expansion, instead of slowing down as would be expected, the rate itself is increasing. In 1998 the observed increase in red-shift indicated that not only was *expansion accelerating*, but the *rate of acceleration was increasing*, a discovery made by two independent observations, one by *Saul Perlmutter* and the other by *Brian Schmidt*. We now have a "run-away universe" that is predicted to expand faster and forever. That is the standard model that is based on two *independent, unrelated observations*; one is the *Doppler effect* for sound waves identified by *Christian Andreas Doppler in 1842*; the second is the observed *frequency down-shift* toward red called the "red-shift" for distant cosmic objects. The first physicist to apply the Doppler effect to stars was *Armand-Hippolyte-Louis Fizeau* in1848. In 1868, *William Huggins* connected the Doppler effect to the red-shift of *Fraunhofer absorption lines* to determine the velocity of stars receding from Earth, which was "confirmed" in 1871 and "verified" by *Aristarkh Belopolsky in* 1901. Working in optics in the early 1800's, *Joseph von Fraunhofer* developed spectroscopy and discovered the dark absorption lines of various elements, relating them to the red-shift. Every chemical element has its own "signature" absorption-line pattern (like its own fingerprint) by which it can be identified from the Light of stars. The specific *pattern* of emission or absorption lines is constant even when the pattern is shifted up or down in wavelengths. In **Figure 29**, the color labels at right indicate the drop of the Fraunhofer spectral lines from the left to right columns as a frequency down-shifts toward the red. The amount of shift up or down is assumed to be a measure of the distance as well as the velocity of the emitting object, but *only* if the object's distance is calibrated to some other measure of distance, and *only* if the observed red-shift is *in fact* entirely due to the Doppler effect. Around 1931, *Edwin Hubble* published his red-shift measurements of distant nebulae based on the Doppler effect, correlating and calibrating the red-shifts with pre-established distances as measured by what are called "standard candle brightnesses*" intensities*" *of Cepheid variable* stars in our Milky-Way Galaxy, or of *Super-Novae and Quasars* for deep-space objects.

These two independent phenomena, *the Doppler effect and the cosmic red-shift,* have been assumed to be linked together beginning in 1848 and repeated in 1868, 1871, 1901, 1912, 1923 and in 1931, the 1848 setting a "precedent" for an assumed connection that gained momentum with each repetition over time, lending a faux-air of "certainty" to that assumption. There is only one experiment that appears to uphold the red-shift-as-Doppler interpretation: different objects at different distances from Earth were recorded over the same interval of time; the more distant objects showed a greater red-shift than those closer to Earth, but the red-shifts were well ordered according to their distances.[42] Despite my bias, the red-shift-as-Doppler may still be factual. I suspect that the spacings of the Fraunhofer spectral-lines in Figure 29 might represent a kind of harmonic sequence of some fundamental vibrational frequency. In 1897, *Pieter Zeeman* demonstrated that a *magnetic field would separate the Fraunhofer lines into several variously spaced lines* (the *Zeeman effect*), and in 1913, *Johannes Stark* showed that the same kind of splitting of the lines also occurred in *electric force-fields*.[43] The Zeeman effect of using electric or magnetic fields to split the Fraunhofer lines for various elements appears to be related to the *harmonic multiples of a fundamental frequency*, which is what I believe the "quantum" to be.

Einstein's equations suggested an "unstable" universe, one that either expanded or contracted. However, Einstein had a personal bias for a "stable" universe, so he manipulated his own calculations by adding a figure he called the *"cosmological constant"* that made the universe "stable". When he heard of Hubble's red-shift findings that the universe seemed to be expanding, Einstein promptly and without any question removed his cosmological constant, calling its insertion "the biggest blunder of my career." Shortly after Hubble announced his findings, however, *Fritz Zwicky* proposed his so-called "tired light" hypothesis,* but it was never tested, and after the discovery of the *Cosmic Microwave Background* (CMB) in 1964 by *Arno Penzias* and *Robert Wilson*, Zwicky's idea was laid to rest in favor of the Doppler-expansion model.[44] Zwicky had proposed that even light must expend energy in its travels, partly as a diminution of brightness (intensity) over distance, but also as a loss of energy by a *lowering of frequency*, so that an increasing red-shift reflected *not a universal recession-expansion of space-time*, but only the *distance light had traveled* to Earth from its far away source. Regarding the CMB, there are other interpretations to be discussed later, along with a "stable / unstable" universe. In 1923, *Alexander Friedman*, and later *Georges Lemaitre* in 1927, independently derived indications of a "non-static" universe from Einstein's equations. However, a question arises as to whether "unstable" *necessarily* means either "expanding" or "contracting", or if "unstable" or non-static" might simply mean *dynamic* in some other way (This will be discussed later).

There are problems with a "runaway" universe besides the absurdity of expansion forever. Assuming for a moment that an accelerating rate of expansion is a given, what would happen as that expansion speed approached the speed of light? Unless Einstein is wrong, it is current doctrine that the speed of light at 186,282 miles per second in vacuum is an absolute speed limit and that nothing can go faster than that.[45] According to Relativity, since *length* is measured in the direction of motion-expansion, *lengths contract to zero and time slows to a stop*. Not only expansion would stop, but space—at least as expanding "length"—would vanish. Either the evolving universe would cease altogether, or at least expansion would stop and universe would be without any movement at all. That would mean that *Energy as work-action would have no* "capacity to do work", which is a *violation of the law of conservation* that was proven by *Emmy Noether* (as cited at the beginning of this work).

Black holes are at the center of almost, if not all, barred and spiral galaxies (and perhaps even other kinds), and galaxies do sometimes "collide", so another scenario might be a coalescence of black holes into one immense singularity of infinite mass density and infinite temperature as a prelude to another Big bang and another evolving universe and so on. Either of these might give rise to a recurring cycle, but then there are questions regarding entropy as "the arrow of time"; would entropy "accumulate over" each cycle, or would entropy somehow be "reset to zero" at each new singularity?[46] Speculations about what might or might not be that cannot be tested are moot and a waste of time. As Gotama *Siddartha* is said to have explained to a disciple wanting to know about the universe, *speculation* is but a "wilderness of opinion."

Since the time of Hubble, physicists have realized that there are several other causes of red-shift besides Doppler; 1) *Relativistic red-shift* is due to the differential motions of two or more frames of reference relative to each other that involve time dilations and length contractions; 2) *Gravitational red-shift* (also called "Einstein shift") where a clock ticks slower deep inside a *gravitational* "well" as measured from outside the g-well so that the *outside measure exhibits a red-shift coming from inside the gravity-well;*[47]" 3) A "cosmological" red-shift is due to the stretching of space-time itself as the universe expands, but that "expansion" is still based on the assumption of a Doppler effect; 4) *Frtiz Zwicky's* "tired-light" *is* not a different red-shift like the first three, but rather a different *interpretation* than the Doppler model of the red-shift, one based strictly on the *distance Light has traveled.* Zwicky also claimed that most of the mass of the universe is invisible, which later led to the idea of "dark matter" and ultimately of "dark energy". But *no one has ever seen Energy itself*; Energy is only experienced as its *manifestations of gravity* (which also no one has ever seen), *radiation, and as matter in motion*, ie. as *kinetic energy (*EK). The unmanifest EWF is itself invisible and therefore "dark", so what need is there of another special "dark energy" to explain *why* the universe "falls outward" (as one pundit put it.)?

The "*Cosmological Principle*" (CP) is another part of the "standard model", first published by *Isaac Newton* in 1687. It maintains that the distribution of matter in the universe is *homogenous* and *isotropic* when viewed on a large enough scale. "Isotropic" means that the universe is the same in any direction for any observer; the "perfect CP" states that universe is not only homogenous and isotropic in *space, but also in time*, that it always has and always will look the same. It is also implied that a universe that follows the CP must be non-static. "Experts had suggested that the maximum size for any cosmic structure would be about 370 *Mega parsecs*, (Mpc) where *Mega* means millions, and a *parsec is 3.26 light-years, or 19 trillion miles*. However, in 1991 and 2011, *Large Quasar Groups* (LQG) had been measured at lengths of 580 and 780 Mpc respectively,[48] and in 2013 a new structure 10 billion light-years away was found that measured 2000 to 3000 Mpc. The largest known "structure" in the universe was found in 2004, a gigantic "supervoid" measuring a *billion light-years* across.[49] It is not entirely void of mass, but the distribution is much thinner than elsewhere. It seems that at a large enough scale and at greater distances, the universe appears to be considerably *less isotropic and homogenous* than the CP asserts, with hugh clumps and walls of galactic groups, and at least one gigantic void-like region, and there are most likely others to be found. All of this calls into question the assumptions that have been made and that are still being taught even in the face of new findings. The "cosmological principle" was criticized by *Karl Popper* on the ground that *it makes our lack of knowledge a principle of knowing something.*

Gravity as an "attractive force" seems to hold the universe together. Estimating the total *gravitational mass-density in the Universe* disregards two points: The first is that all emissions from Light sources (and reflections from non-emitting objects) are *visible and measurable only by line-of-sight* (LOS).[50] Electro-Magnetic energy (Light) is radiating from stars in all directions, yet *between* stars, the night sky is dark, even though "filled with waves of Light". It means that *Light is invisible to us except and only by direct line-of-sight* (unless also scattered by dust or gas). One explanation given is that all the varying wavelengths cancel each other out, but those wavelengths reaching us by LOS would also be "canceled out", so that explanation doesn't hold. Secondly, it means that the major amount of EM radiation is not counted, or is only estimated as gravitational mass-density, which is accepted as what appears to be an "insufficiency" of mass-density to hold the universe together, so "it falls outward" as expansion. It has been proposed that a "dark matter" makes up about 70% of the material universe, yet this still is claimed to be "insufficient" mass. Not only is dark matter *invisible*, but it is apparently *undetectable by any chemical analysis and neither is there any detectable "charge", emission or reflection of Light*; it is only *inferred as a gravitational effect* on observable "normal" material objects. "Dark energy" on the other hand, is a purely speculative idea, a strange kind of "negative and repulsive" gravitational energy that makes the universe "fall outward" instead of holding together.[51] It is often claimed that this dark energy has been discovered. *It has not.* What *has been* discovered in 1947-1948 by *Hendrik Casimir and Dirk Polder* is the "*Casimir effect*", a force between the atoms and molecules of two conductive but *uncharged* metal plates. *Diderik van der Waals* (1837-1923) had previously discovered the force between *intermolecular, electro-static* forces that existed between two parallel uncharged metal plates, now known as the "*Van der Waals' force*". Measurements of that force by Casimir in 1955 found it to be strongest for plates at extremely close proximity that weakens rapidly with distance. *Neils Bohr* suggested that the Casimir effect might have something to do with "zero-point energy" as a sort-of "negative mass=negative gravity", also known as "vacuum energy", aka "dark energy". There seems to be a kind of desperation to leap on anything that might clarify or mystify Quantum Theory. *Schwinger, De Radd and Milton* in 1978, and *Robet Jaffe* independently in 2005 published statements that "*the Casimir effect can be formulated and computed entirely without any reference whatsoever to any mysterious zero-point energy.*"[52] In point of fact, since no one has ever seen or touched *Energy itself, all* Energy is "dark", and any contrivance to explain a "runaway universe" seems—well, contrived. If the universe is in fact "flying apart", there must be some other explanation not yet discovered. But perhaps the red-shift has been *misinterpreted*, so there'd be no need to explain an "accelerating rate of expansion". With variables of size and distance using assumed "standard candles", there may not be any good way to determine the cause(s) of observed cosmic red-shifts,[53] but Zwicky's and "gravitational red-shift" seem to be the best interpretations.

Systems

According to texts, there are four types of systems that define how "permeable" their boundaries are to different forms of energy. Although "boundary" implies an inside and an outside, an *Isolated system* technically has no "outside" since all energy exchanges happen entirely *within* the system. There could only be one *Isolated System* in the universe, and that is the *Universe itself.* The opposite to "isolated" is an *Open system* where the boundary between inside and outside is entirely "permeable" to any and all energy exchanges. Technically, *all systems within The System* are "open", except *perhaps* in "controlled laboratory conditions". A system that is completely open basically *has no distinct boundary*, so "inside" and "outside" do not seem applicable. A system that is "isolated" has no "boundary" either, there being no "outside", and even labelling the "inside" of an *Isolated system* seems to be a moot point. The other two systems are defined by the boundary's limited openness to this or that *form* of energy. For example, a *Closed system allows only energy itself to cross the system's boundary*, but prohibits *matter* from entering or leaving. Since Energy is suggested to *overflow the manifest Universe, Closed* might be more appropriate than *Isolated.* An *Adiabatic system* however, *prohibits* "heat" from crossing its boundary.[54] "Heat" however, is just one form of Energy. Both "Open" and "Closed" can apply to ordinary "subsystems", and "Adiabatic" would be applicable *only to* sub-systems *and not* to the Universal System of all systems. Heat certainly transfers across system boundaries in all kinds of actions and reactions, which is why it is maintained in this paper that our Universe is *by definition a Closed or Isolated System.* If Universe is indeed expanding, it is not into any pre-existing space; it would create its own space by its very expansion, so there'd be no "boundary crossing" at all. But neither could expansion create new Energy, nor can Energy be said to "thin out" with expansion. The Adiabatic system seems somewhat contrived in order to allow all "waste heat" (from friction, etc) to accumulate "inside" the universe as increasing entropy to a point of thermal equilibrium where all temperature is the same. *Without heat-flow, nothing can happen anymore.* In keeping with energy conservation, it is said (begging the question) that "energy is still present; its just not available to do any work". But the very *definition* of "energy" is *the capacity or capability to do work.* "Unavailable" means *zero capacity or capability to DO work,* ergo—zero energy, a clear violation of the law of conservation. But wait. If there is any temperature present above zero Kelvin, that means that some energy is present. In an Adiabatic system that keeps heat inside that finally reaches thermal equilibrium, *energy is both present and not present*, which is why an Adiabatic system applied to Universe is self-contradictory. It may have been invented specifically to accomodate the idea of "waste heat" accumulation to entropic equilibrium, and to "explain" space-time expansion. In mechanics and chemistry, heat is always crossing boundaries. But any system can be theoretically narrowed or expanded to exclude or include whatever seems applicable to a specific situation. The term "system" is relative to variable parameters.

The Cosmic Microwave Background (CMB)

In 1940, George Gamow predicted the existence of a "background radiation" from shortly after the Big Bang, and in 1964, *Arno Penzias and Robert Wilson* detected a low frequency background that was interpreted to be *a remainder of— and verification for*—the Big Bang beginning of the universe and it's subsequent expansion. The temperature today is almost 3 Kelvin, just slightly above absolute zero and is inferred to come from around 379,000 years after the BB when temperature is estimated to have been near 4000 Kelvin.[55] But there are other possible interpretations for this background radiation. Internally, stars are continually creating atoms and molecules of Hydrogen, Helium, Carbon, Oxygen and other various elements, and this production is suggested as a possible source for the CMB. Another idea is that the CMB is a background Electro-Magnetic noise from spinning black holes as well as the EM energy from an "all-pervasive sea of unmanifest zero-point or vacuum energy". "Spin" also seems to create a "polarizing radiation effect" along the axis of spin that might account for the low frequency background. But more to the point is that the CMB is held up as *both the result of—and simultaneous proof* of the "Big Bang", putting to rest Zwicky's and all other non-Doppler interpretations of the observed red-shift, and the big "CMB".

A glaring anomaly concerns the CMB and time-distance. The universe is currently estimated to be about 13.7 billion years old, and the CMB is thought to be around 13.4 billion years old. The CMB is said to exhibit "an enormous red-shift*"* due to "wavelength stretching from space-time expansion". The temperature at 4000 Kelvin from 13.4 billion years ago has diminished to about 2.7 Kelvin. But the detection by the Chandra X-ray telescope of a very large (gigantic by our standards) X-ray emission source around 12 billion years distant (*Galaxy Cluster 4C41.17 near Auriga*),[56] evokes the question "Why is the CMB so strongly red-shifted when emissions from an X-ray source only 1.4 billion years later are not? The difference of 1.4 billion years is sufficient time for even *greater space-time* expansion (according to the Doppler model), *so there should be a greater red-shift for those X-rays*, or at least *similar* to the CMB. Perhaps the x-ray's initial emissions are of an unimaginably higher off-the-charts frequency; but what would be higher than 1.4 billion years closer to the BB? During that interim, expansion should have increased more than during the Cosmic Microwaves; yet those Auriga emissions detected were still apparently at X-ray frequencies. This heightens the discrepancy regarding the CMB as "verification and proof" of the Big Bang expansion model of space-time and the whole standard cosmological model. Add to these anomalies the variations of gravitational curvatures and the bending of light that is observed as *line-of-sight* trajectories, and the standard model is very shaky.

Spin and Polarization

What makes everything *spin* in the vacuum space of the macro-scale universe or in the micro-scale of "quantum particles"? Would there be two different causes, or would spin at all scales be an effect from one cause? I was told by a physicist that the individual micro-spins of particles gradually became clumped together by gravity, with their spins—like flocks of birds or schools of fish—coordinating into one macro-spin. But that didn't explain the micro-spin of individual particles. Where does spin come from? It has been pointed out that the choice of the word "spin was unfortunate" because there is "no literal spin in the quantum world". The term refers only to some "characteristic of a quantum particle" that someone called "spin".[*] Nevertheless, at least at the macro scale of the cosmos all spheroidal bodies do actually spin; even asteroids and space debris seem to rotate while haphazardly and randomly tumbling through space-time. But even the so-called "energy-particle photon" *has angular momentum, so it must have spin, and having spin, it must have mass.* However, it is misleading to talk of "spin up" and "spin down", because they are "observer dependent" (see **Figure 30**)[57]

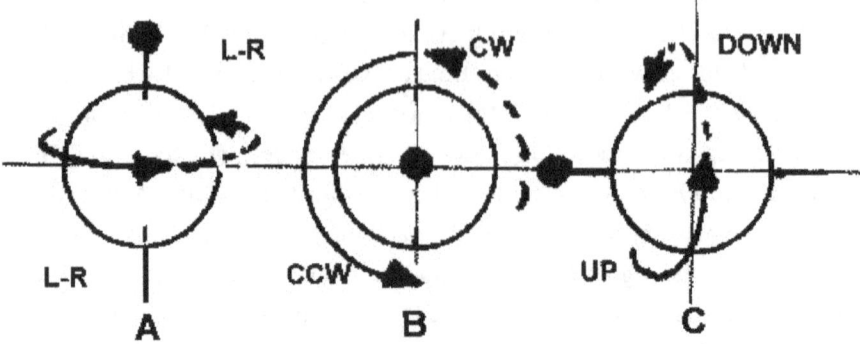

Figure 30 Spin A is observer independent. Spins B and C are both observer dependent.

Between three possible spin axes, there is symmetry as "observer independent" *only* on *the vertical axis with spin "Left-to-Right" or "Right-to-Left."* The observer (b) in back will see the *same spin as the observer* (a) in front. (in **Figure 30 A,B and C**, below, the dashed lines show the spin as seen by observer "b" on the opposite side of observer "a.") This is due to the bilateral symmetry of most vertebrate life forms on this planet. Let's say the horizontal spin is left-to-right as in **30 A**, and that will be the same for both observers. Now swing the axis halfway down so the top is directly in front of you as in **30 B** so that observer "a" sees spin as "*counter-clockwise*" while observer "b" sees that *same spin* as "*clockwise'*. Turning the axis amounts to the same thing as changing the observer's positions relative to the axis-of-spin. *Now turn the axis again horizontally* so that the top of the previously "in-line" horizontal axis is now to your left, and the first spin in 30 A "L-R" for observer "a" and "b") is now "*up*" for observer "a", but observer ("b") sees that same spin as "*down*" (as in **30 C**). Both these last two spins on the two *horizontal axes* are "*observer dependent*" (ie. *the spin direction depends on the observer's position relative to each horizontal axis).*Given the three dimensions, it is obvious that there must be two of one and one of the other, but there is *always one vertical axis and two horizontal axes*, and never the otherway around. But if those same three axes are the center axes of *planes*, the planes are *always and only two* vertical and one horizontal planes on *one* vertical and two horizontal linear axes. *Only the spins "L-R" or "R-L"* are *independent* of the observers' opposite positions, *but they are all—in 30 A, B and C—exactly the same spin!* In "quantum-speak", the designations "spin up" and "spin down" are nonsense since spin up and spin down are just observer dependent and different for two oppositely positioned observers. A "*relativity*" and not a "probability" in the "quantum world"?[58]

A spinning top or any spinning sphere will travel in a curved trajectory, apparently even without the effect of gravity or *Coriolis effect*, but by its own angular momentum. A side but related question is: "How can this geodesically curved trajectory be distinguished from the curved path imparted by gravitational mass? Or is there any distinction to be made? *The spin of a sphere traveling on a curved trajectory will trace a spiralling movement* (See Figure 26, page 19). Fragments from colliding particles in accelerators will spiral and coil in a magnetic field. Could the *unmanifest Energy Waves* themselves "twist, bend and coil" so as to come into self-relational interference that—as pattern and form—*is equivalent to curvature and spin*?

The mathematician *Kurt Gödel* suggested to Einstein that perhaps the whole universe spins or rotates, with spin being the "geometry of space in which time disappears."[59] In the word "universe", *uni* means "one", "single" or "unity", and *verse* comes from the Latin *vertere* meaning "to turn" or "turning", so Gödel just might have something, especially the idea that "time disappears". If the whole universe would turn or rotate, that could impart spin to almost everything. The spin of black holes at the center of galaxies impart a "frame-dragging" spin to the stars, dust and gasses that become the galaxy's twining arms; but what imparts spin to a black hole itself? Is the self-acting action of an Energy-wave itself spin or spiral?

Einstein explained *Brownian motion*, the observable random motion of insoluble particles suspended in a liquid as "colliding with the atoms or molecules of the liquid itself". Such collisions are probably not all "head-on"; oblique "bumps" could impart a spin to the particles, but there would be no over-all coherent and coordinated motion as in schools of fish. In the *Energy Wave Field hypothesis*, "matter" is a *spheric wave-form*, and as a wave-form it could spin *as though it* was a "particle". But *what causes spin in the first place* is still unanswered, and at this point there is as yet no satisfactory explanation.

There is a scientific toy called a "radiometer" with a four-bladed vane delicately balanced on a swivel, all housed in a glass globe. Each vane is silver or white on one side and black on the opposite side, so that when light or heat is shone on the globe, the white side of each vane reflects the energy back and away while the black sides *absorb* the energy and makes the "light mill" turn the same way that wind makes a wind mill turn. All incandescent light generates heat as well as light waves, and a test shows that *heat without light* will cause the mill to turn, but a much cooler LED light will have little to no effect on the "light" mill. It appears that it is more of a "heat mill", since a cooler LED light has little effect. There seems to be much more to light than meets the eye, besides what we only see at line-of-sight.

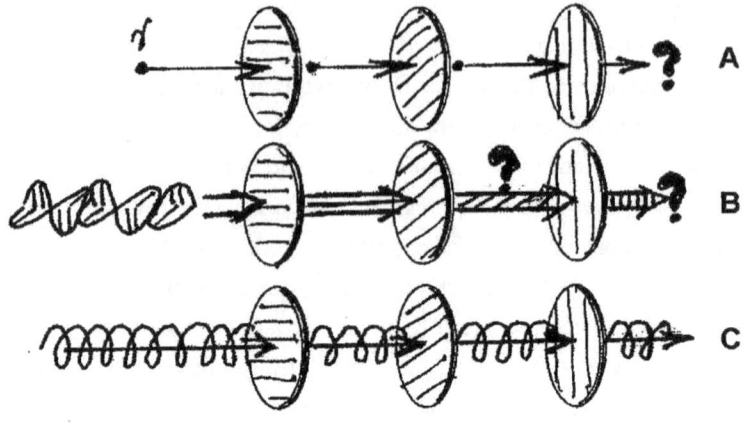

Figure 31 Polariztion of A-photon particle; B-perpendicular Electro-Magnetic wave; C-EM wave as a spiral

In Quantum mechanics, *polarization* of a "particle "is as mysterious as a particle in the two-slit diffraction experiment. Polarization of a "*particle*" requires *angular momentum*, the direction and speed of spin relative to the particle's linear motion, the direction and degree of any oscillation, and a quantum number for spin that happens to be different for "*boson*" and for *fermion* particles. Complicated? Another question is "What constitutes the *oscillation* of a spinning particle? Is one complete "spin" an "oscillation", or can a particle oscillate—ie. *vibrate*— while also spinning? I can see that explaining the *polarization of a particle itself* would be complicated, although a particle's "oscillations" could very easily be explained. (**Figure 31 A**) In Maxwell's wave theory, the *electric and the magnetic wave* components each oscillate on a plane at right angles to each other, both planes also being perpendicular to the direction of propagation (see **Figure 2** and **Figure 31 B**). This transverse wave-image makes the polarization of light easier to understand, but it appears even polarization of light *waves* is strange in that wave polarization can also be *diagonal* and not just vertical and horizontal. For example, in **Figure 31 C** light is passed through a *horizontal filter* allowing passage only of waves on the horizontal plane; the polarized light-beam then passes through a *second filter at 45 degrees oblique*, and then through a *vertical filter* that *surprisingly allows vertically polarized light to pass through*. How can that be explained by right-angled EM planes as in B, even in wave theory? It suggests other waves at 45 degrees or quite possibly waves *propagating in a spiralling motion* as in Figure 31 C, *rather than a simple linear movement.*[60] Passing a spiralling wave through the two-slits of a diffraction experiment would polarize through any angled oblique filters. One question would be "what is the effect when using oblique polarizing filters *at other than 45 degrees?*" (The twisting and kinking of electric cables and cords might be due to a spiralling effect of current moving through the conductors.)

Gravity

Except for the effects on physical bodies, gravity is invisible. It is inferred and named to explain those effects. Even with Einstein's General Relativity, Gravity has yet to be understood; yet there is a "race to quantize gravity" in search of a "Grand Unified Theory" (GUT). Einstein posed that Gravity is equivalent to acceleration, which is any change of speed and / or direction. What was once called "centrifugal force" is an example of acceleration as gravity. *All curved trajectories are accelerations related to that of a spinning sphere, so spin and gravity seem to be related,* but I cannot as yet see or say how. Spin might also be related to "charge".

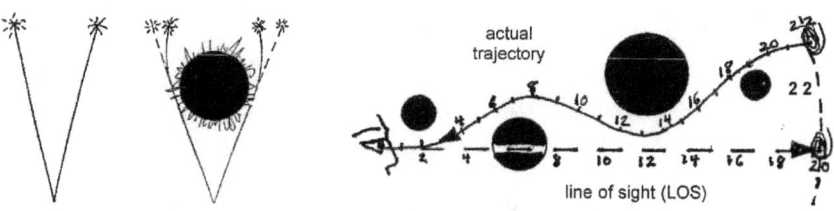

A Star positions before eclipse   **B** Star positions during eclipse observed as line-of-sight

C Light's longer trajectory curved by gravitating masses compared to observed shorter line-of-sight

Figure 32 Eddington's experiment for gravitational bending of light

Figure 32 C Geodesics from gravitatiional fields

When Einstein's equations suggested that gravity would bend light-rays, the test for that was possible only when a complete solar eclipse was predicted to be visible on May 29, 1919 from an island off the west coast of Africa. *Sir Arthur Eddington* traveled there to measure the distance between two stars *prior* to the eclipse (**Figure 32 A**), then measured again *during* the eclipse and found the stars to be "farther apart". The discrepancy was due to the sun's gravity bending the star's light rays, but what was observed and measured was by *straight line-of-sight* (the dashed line in **Figure 32 B**). It is obvious that in *Euclidian* geometry, a curved line is longer than a straight line, so if our sun bends starlight that much, what happens to Light passing by more massive stars or traveling over millions and billions of light-years of uncountable gravitational fields of immeasurable strengths (**Figure 32 C**)? According to Einstein, "the shortest possible distance between two points" of a light trajectory is a "geodesic" curve due to the gravitational curvature of space-time itself. Seeing and measuring Light *only by direct line-of-sight* makes observed distances, velocities, red-shift calibrations and even objects' *positions full of error*. Looking out into the vast distance of the Universe, nothing is as it seems. A too-short line of sight distance relative to a longer actual curved trajectory means that greater "red-shift-as-Doppler" would be interpreted as an "increased recession-expansion", and it would be wrong.

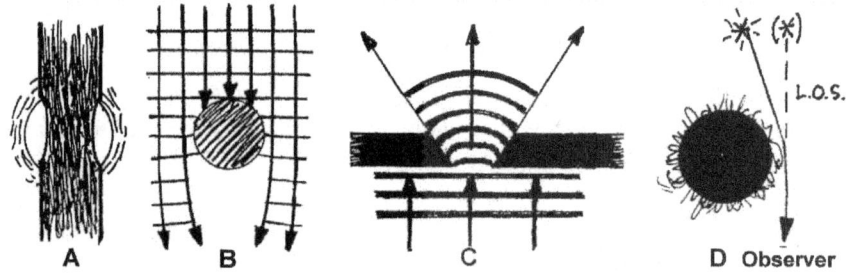

Figure 33 Light diffraction as gravitational bending?

A purely hypothetical question concerns the relationship between the gravitational bending of light by the sun, (and according to Einstein, by all matter) and the bending of light by diffraction. **Figure 33 A** illustrates how light from the sun behind an object like a tree or post will bend light around and partially "negate" the form. This could be merely an optical effect of the eye's lens, or it could be related to matter's gravitational bending of light. **Figure 33 B** is a top-down diagram of light waves as they bend around the object toward the observer. **Figure 33 C** is a close-up of a *rectilinear* wave-train passing through a small aperture as though the "material edges" of the aperture exert a sort of "drag" or "pull" on Light-waves as they "diffract" from a straight into a *curvilinear* wave-train that radiates outward. **Figure 33 D** shows light bent by sun's gravity. The question posed by these figures *is whether or not the diffracrion of Light by matter might be equivalent to* *gravitational bending?* The other "ordinary" light phenomena of reflection and re-fraction as changes of direction and speed, are also *accelerations* that—according to Einstein—are equivalent to gravity. (Both "reflection" and "refraction" presuppose the existence of 'matter" as spheric wave-forms, as does "diffraction".)

Newton's Sleight-of-mind Gravity Trick

As taught in all school textbooks, Newton's thought experiment moves an object from a gravitational source out to infinity where gravitational strength diminishes to zero, and the potential energy (Ep) of position is then also zero. According to Newton's terms, man's work is "positive" against the "negative force" of nature's gravity. For a science whose goal is to "understand nature", *this is a blatant anthropocentric bias.* But that is not the "trick". Man's *positive work-energy* as "force" in moving the object out to infinity is transformed into *kinetic energy* (Ek), and as gravitational strength (g) diminishes with distance, energy input as "work-force" *against* gravity also diminishes; but the total Ek is still nevertheless *cumulative.* Gravitational *potential energy* is the "stored energy of position" where gravity still exerts some force to "pull" the object back to the g-source. The Ek of motion (by "Newton's work") would be transformed into Ep of position at Newton's "infinity". The question—and the trick—is: if Ep is "0" at infinity, *what happened to that cumulative work input as Ek motion if the final Ep is "0"?* All that cumulative "work-energy as motion" has *disappeared, meaning "energy has been negated",* a blatant violation of the principle of energy conservation. A deeper question is the contradiction in the assumptions themselves. For example, the Ep as zero "elastic" g-force at infinity transforms (ie. negates) all cumulative Ek to "0" Ep at infinity. Gravity is zero, and Ek is positive, but transforms to Ep as zero! Something is fundamentally wrong with our view of nature and her "forces" to yield such a contradiction, and *this is still being taught in schools as the basis for the negative gravitational calculations in use today.* But "+" and "- -" assignations aside, *the equations still seem to work for moon and Mars shots.*

What is "work", or more specifically, what is "useful" work? Gravity, whether a "natural force" or an "architectural geometry of space-time" is "attractive" and pulls or holds together the objects of the Universe. That is its "work" and would that not be more "useful" than man's opposition to that? Gravity—whatever it is—has been around long before Earth was formed, even before our sun was formed.

Is our "human work" more "useful"? More than one textbook states that "the downward direction of gravity in free-fall *should be considered as positive*," *and the effect is at maximum at the surface of the mass*. If this is the case for free-fall, should it not be the case for all gravitation, reversing Newton's gravity as a "negative force" against which humans exert "work" in opposition? The acceleration during free-fall is due to the increasing gravitational strength approaching the g-source but at the center of a massive black hole, *gravitation is zero*, since all of a body's mass is between the center of gravity and the surface of the body's mass. *Counting gravity in equations as negative changes the whole valuation of energy as mass and skews the estimate of "mass-density"*. Is it possible for a "positive mass" to effect a "negative gravitation" when all Energy is mass and all mass is gravity? When "attraction" or "holding together" are held to be "negative", what effect might that have in human society? Science, erroneously viewed as "fact" is not divorced from all human culture and its values. All cultural forms—social, economic, political, mathematical, scientific and philosophical—are inextricably inter-related. Reversing sign valuations so that gravity is "positive" and human work in opposition is "negative" might not affect the numerical calculations *except for the reversal of signs, but what effect might a reversal of signs have regarding the runaway expansion of the Universe?* Changing signatures is not without precedent (Penrose *Cycles*…,pp.89-90). Newton's sign valuations are glaringly homo-centric, based on humans *being in opposition to* nature and reality. Reversing the signatures of gravitation has huge ramifications.

Position and Motion in Deep Space
A parallax method is believed to be the most accurate for estimating cosmic distance, size and motion. The trigonometic calculations involved are said to yield an *accuracy within a 2% margin of error*.[61] Distance by parallax is then calibrated with the assumed Doppler red-shift to yield the "speed of cosmic recession-expansion". Some details of estimations within that 2% need to be examined in detail. That small margin of error when projected into deep space of millions or billions of light-years becomes enormous in terms of estimating brightness, distance, size and even position. For example, using our familiar *miles per hour* as units of measure most familiar to non-scientists, a *light-year is 186,282 miles-per-second times 60 seconds per-minute =11,176,920 miles per minute, times 60 min/hour =*

670,615,200 miles/hour, times 24 hours-per-Earth-day= 1.60947648 x10 10 miles *a day, times 365 days per "Earth-year=5.874589152x1012 miles a year as the* *distance light travels in just one light-year. That is 5,874,589,152,000 miles/year* *or almost* <u>*6 trillion miles*</u> *in just one Earth-year. This is now multiplied by .02 error* *= 1.17491783x1011= 117,491,783,000, or over* <u>*117 billion miles error*</u> *for every* *single light-year interval!* Multiply this by the millions or billions of light-years of deep-space distances, and this is just a simple linear progression of a 2% error margin, not including the exponential growth of small errors or omissions of data that can produce anomalies or the sudden onset of unpredictable chaotic behavior of our so-called "knowledge" of deep-space systems. Now add to this the unknown and un-knowable multiple geodesic trajectories of Light traveling by and through untold gravitational fields, (Figure 32 C) plus the slowing of light-speed through media such as gas and dust clouds, and just what is it we pretend to measure? Even *positions* will be wrong due to a line-of-sight measure of Light's curved geodesic trajectories from deep-space to our instruments. This is not counting the updated changes due to new technologies and more precise measurements that are certain to be made.

British astronomers have discovered a huge "Large Quasar Group" (LQG) measuring 4 billion light-years across (our Milky Way galaxy is tiny in comparison at about 100,000light-years across). But the largest structure is a gigantic void where few galaxies exist (about 20% less dense than any typical region in the universe), discovered when astronomers investigated an unusually large, cold region detected in 2004. A team headed by *Wendy Freeman* of the *University of Chicago* discovered a 9% discrepancy between observations and the "Hubble constant"[62] (which is assumed to describe the speed of expansion of the universe) Another example of error is the estimated distance from *Earth to the Galaxy NGC 2841 in the Ursa Major constellation*, estimated in the year 2000 to be about 30 million light-years away; observations in 2010 of the galaxy's *Cepheid stars* corrected that distance to 46 million light-years, a difference of 16 million light-years in just a ten-year interval. Compared to billions of light-years distances, 46 million is relatively close to Earth. The *under-estimation in 2000 would make any red-shift seem greater than should be* for the closer distance estimate of 30 million and *would be Doppler-interpreted as accelerating expansion* and it would be wrong. Science is inherently an ongoing and unfinished enterprise, but the accumulation of claims and ad hoc adjustments to *inertially* sustain an unsustainable "standard model" in the face of new data is counter to the scientific endeavor "to know".

The relative motions of two gravitational bodies are fairly easy to calculate since the days of Isaac Newton, but add one small third body and in theory and practice and the complexity is increased exponentially. When attempting to work out this "three-body-problem" (as it is called), the mathematician *Henri Poincare* encountered his "nightmare of chaos" that eventually led to Chaos theory, known popularly as "the butterfly effect". Only by forced computer simulations has the chaos been visually exposed, but even then the computer can go only so far before uncertainties take over. However, working from the earlier works of *Karl Sundman, D.G. Saari, Q.D. Wang, and Z.J. Xia* generalized from three bodies to a solution for an "n-body problem", but the calculations were so complex as to be useless for predictions over long intervals of time.[63] The complexity of just three bodies is difficult enough itself, but groups of bodies in globular clusters make calculations impossible. Imagine trying to "solve" the movement of every particle in "Brownian motion"! Small "insignificant" errors or intentional simplifying by superposition or by dropping a few decimal places make errors grow exponentially until "chaos" sets in and prediction becomes impossible. Chaos theory however, is not merely an artifact of measurements or calculations; *periodic systems* seem to require a brief moment of "chaos" as a way to "reset" the periodicity of a system's behavior that otherwise would descend into non-periodic, unpredictable, random and turbulent fluctuations. There is still much to discover about the "hidden order" in *Chaos*, as well as the "mysteries" of Light and gravity.

Gravity, Spin and Space-Time

The black holes as "spinning nuclei" of galaxies impart their spin to the galactic stars, dust and gas clouds in a "wrap-around" effect that forms galactic "arms", and perhaps impart spin to the stars, planets and satellites themselves. In 1934, *Fritz Zwicky*, and in 1975, *Rubin and Ford* independently observed that the velocity of stars rotating around the galactic edges appear to be moving faster than predicted according to observed estimations of the distribution of "normal" matter. "Dark matter"[64] has been proposed to explain the discrepancy, but other gravitational anomalies have been accumulating. The Doppler interpretation of the red-shift observed in 1998 as an "accelerated rate of expansion" is alternately explained as an effect of the *non-uniformity of gravitational strengths, with differential space-time curvatures throughout the universe.* Another anomaly within the standard model is that EM radiation waves traveling through galactic clusters should first gain when entering—then lose energy when leaving the cluster; but the radiation of the CMB seems to gain *twice* as much energy as predicted, suggesting that g-strength weakens *faster than the inverse-square law.* On the other hand, according to spectral lines, hydrogen clouds appear to "clump more" than expected, which indicates that g-effect weakens *slower* than the *inverse square law.*[65] It is obvious that estimates of gravitational strengths are considerably more varied than any assumed singular smooth curvature for all space-time (Figure 34 A).

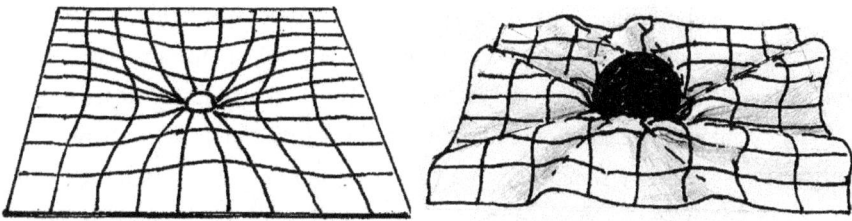

Figure 34 A & B Gravitating mass distortions of a 2-dimensional space-time grid
A Gravity as usually depicted
B Space-time distortion of 2-D fabric.

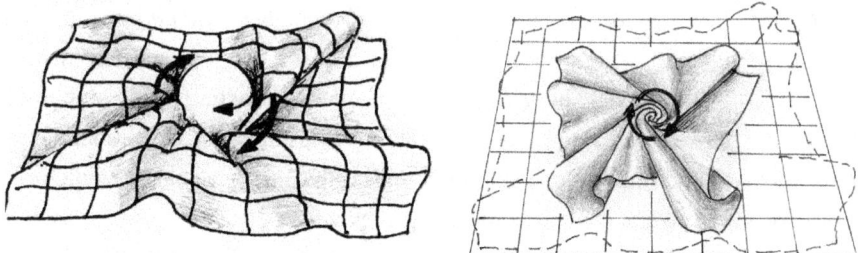

Figure 34 Frame-dragging twist and contraction from a spinning gravitational mass
C Space-time twist from a spinning mass
D Space-time contraction from continued spin

As mass, both EM radiating waves and matter-waves have their own effects on any gravitational curvature. It appears as though all spheroids spin to some degree, and the more massive the greater the frame-dragging "stress-twist" to the surrounding "fabric" of space-time. Even a non-spinning heavy mass in Einstein's gravitational theory may not impart a smooth warping of space-time, but might cause 4-D "stress-wrinkles" (**34 B**), as well adding the frame-dragging twist, with space-time more complicated than supposed (**34 C**). As depicted in 2-D, these "warpages" actually would be in 4-D cosmic space-time. Representations rarely go beyond an initial "frame-dragging" twist of space-time by a black hole but in **Figure 34 D**, the same kind of metric used in C is used in D (just tilted a bit more). The dashed line shows the "fabric" of space-time from C; as the black hole keeps spinning and gaining mass and angular momentum (ie. accleration equivalence), its surrounding space-time not only twists into contorted folds, but 4-Dimensional space-time is actually *drawn in tighter and actually shrinks. As space contracts, time slows.* When other black holes at the center of spiral galaxies are doing the same, eventually the local space-time contractions would be drawn together to co-alesce into a singularity of infinite mass-density as precursor to another BB, evo-lution, contraction-to-singularity and so on. Think of black holes not as "things", but as *"wells" of pure gravitational mass-Energy* where all manifest energy as frequency-measured radiation and as wave-forms of matter are destroyed, with only pure raw Energy remaining in a "dark" *unmanifest* state (in an unmanifest Energy field). When all space-time contracts so that all black holes are brought together in one massive singularity, the cycle that we now live in is "destructively completed", and the stage is set for a new BB cycle. The entropy that causes phys-icists such theoretic concern is no concern whatsoever since entropy and time apply only to *manifest secondary wave effects as measurable form and pattern. Neither time nor entropy pertain to unmanifest and conserved Energy.* What is shown in Figure 34 D would actually be in three-dimensional contracting space with time dilated (intervals expanded) until all manifest Energy as radiation and matter along with both space and time disappear, and there is only Energy as naked in-the-raw singularity. I believe that could be a mechanism by which a "reversal" of expansion when all manifest forms are "consumed". If all spinning black holes should co-alesce, *would their spins coordinate so that the resulting singularity would itself spin?* Then that could be the *primordial origin* of all spin. The BB itself would be a spinning explosion, imparting spin to every spheric wave-form, (ie. every "spe-cious quasi-particle") and every cosmic body whether solid or gas. But how could there be "spin" at all if space-time itself contracted to nothing? What would "spin"? *Could it be possible that Energy itself—like manifest EM polar energy—would* *have had a Möbius-like self-twisting and curling propagation of waves*—"acceleration" *as intrinsic to* the self-acting action of vibrational Energy-waves themselves—hence *"gravity prior to "matter".?*[66]

In Einstein's *General Relativity* theory, *gravity* is equivalent to acceleration. The multiple nebulae of dust and gas throughout each of the countless galaxies of the universe constitute changes of density from vacuum-space in which the speed of light slows from Einstein's constant (c) to lesser velocities (v). Not only that, but entering and exiting different densities will *refract all EM energy waves to a different speed and direction as well as different wavelength*. Refraction as a change of speed and direction = acceleration.[67] For Einstein, the presence of matter-mass as gravity effects a "geodesic curvature" of Light's trajectory as acceleration through space-time. The question is "Does the trajectory of light actually "curve" from "gravity", or does it change speed and direction abruptly at an accelerating refractive boundary that we perceive as "gravity'? Nevertheless, these boundary changes of mass-densities imply the prior existence of matter as "cause" of the acceleration effect, which would make Einstein right after all, but not quite as he thought. A spinning sphere represents a constant change of direction as angular momentum (and probably a change of speed over long intervals of time). Angular momentum as spin—and also as a satellite's rotation around a gravitating body—are both *"acceleration" as equivalent to gravity* in Einstein's theory. Seeing that *refraction, spin* and *rotation* are all *acceleration as equivalent to gravity*, would these together explain curved trajectories of Light's journey through space-time? But this again calls into question our "knowledge" even of actual positions of cosmic bodies just as the location of a fish underwater is not where it appears to be due to refraction. However, *the supposition in this paper is that an initial gravitational curvature exists <u>prior</u> to any manifestation of spheric matter waves*. A twisting, spiral-like angular momentum-as-acceleration of Energy Waves would effect curvature and would impart spin to any subsequent spheric matter-wave forms. However, matter as *interference densities will further affect gravitational "distortions" of space-time* as detailed above.

THE ENERGY WAVE FIELD

EWF UNMANIFEST

Textbooks state that waves are not Energy per se, but are the directional *transference* of energy (ie. "carriers" of energy). This is certainly true of *physical waves* in water tanks and oceanic waves, but in Maxwell's theory of Electro-Magnetic Waves, the *field* is explained as the effect of a "moving charge", presumably a charged "particle". Since it had been proven that there was no "luminiferous aether" as a *medium* through which waves could propagate, the question posed by Maxwell was "*What waves?*" *Schrödinger*, a proponent of the actuality of waves had asked the same question, "What waves?" Maxwell's wave theory was apparently the first to posit waves traveling through an *immaterial field medium* (See "*Huygen's principle*" on page 16). In the ocean and in water tanks, it is the water *medium* through which waves travel, but the "energy" is *relatively unmanifest* until the drag of a rising sea-bottom—or extremely high winds that cause the top of the wave to over-ride the lower part and "collapse" to explicitly demonstrate the energy-force carried by the wave as a literal "collapse of the wave function".

The unmanifest Energy Wave Field is one in which either no wave interferences occur, or *there are very brief intermittent and evanescent interferences* (akin to *Dirac's "virtual particles"*). This might correspond to the "quantum vaccum state" where space-time is "empty of all particles", but is nevertheless somehow "present", albeit unmanifest and unmeasurable.[68] But Instead of a "*lowest possible* ground-state", an unmanifest EWF "vacuum state" might be an *extremely high* "off-the-charts" energy state, since *almost every successive wave interference involves a "stepping down" of frequency with corresponding longer wavelengths* compared to the initial interfering wave-sets (See Figure 25, page 18). However, *at some angles of interference*, secondary wave effects appear to actually be of *shorter wavelengths and higher frequencies than the initial wave sets, apparently reversing entropy.* The EWF is "ground" only as a hypostatic "source" for all phenomena (somewhat reminiscent of the *Bondi-Gold-Hoyle* "creation field" from which matter is created to fill an "expanding universe"—only in the posited EWF, there is no "expanding universe", as an entirety, and certainly not as "expanding forever". Also, in the EWF, "wave sets" are not from two or more completely separate wave sources, *all waves of the EWF are from one vibratory source* whose wave pulsations twist Möbius-like to bend and fold by which *the fundamental wave self-interferes as though it were two*, with the secondary interference effects manifesting as wave *moiré configurations* as the "ground" for all phenomena. Using computer graphics, (or at least transparent overlays), this should be observed *in motion* to see the *in-out pulsations of the standing spheric wave forms that appear in pairs when one spheric "matter wave"* is in interference with a linear EM energy wave train.

EWF Manifest: The Observed

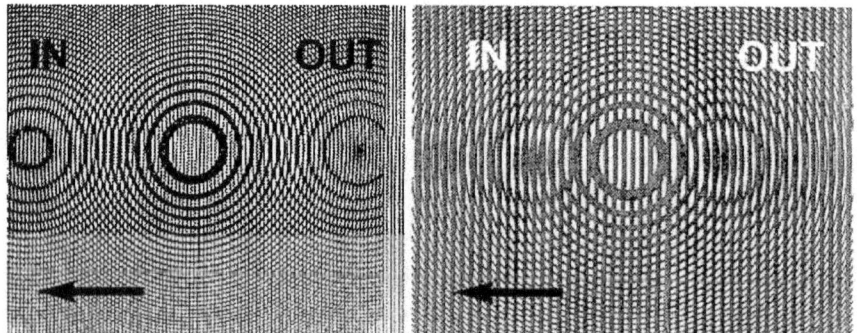

Figure 35 A and B Arrow shows motion of rectilinear waves with secondary spheric moiré waves pulsing in at left, out at right.

Figure 35 A and B

The most significant manifestation of the EWF is when two or more wave sets intersect *at oblique angles*, producing secondary interference wave effects that manifest as moiré configurations. It is also significant that 2-D recti-or curvi-linear wave trains that interfere with a 2-D version of a 3-D spheric wave-form produce *pairs of oppositely pulsing wave-forms, one pulsing inward and one pulsing outward.* Each of these pulsings occur on opposite sides of the initial spheric wave-form (**Figure 35 A and B**; see also Figures 16 and 17 C and D on pages 12-13.). *These are spheric matter waves, but they are not separate from the Field of self-interfering Energy Waves*; they are the same, only *functioning in and as a moiré interference configurations.* In both 35 A and B, when the interfering rectilinear wave-train is moving leftward (arrows), the secondary spheric waves (somtimes there are more than one of each pair as in **B**), *left of the initial primary spheric wave* (center) will pulse *inward as its twin to the right pulses outward.* Reverse the directional motion of the rectilnear train and the twins' pulsations will reverse the left pulsing outward and the right pulsing inward. (With the limitations of physical transparencies, I cannot produce more than this one "fundamental ground-state orbit-shell"; but with several overlays, I can "fake" a nuclear center surrounded by three pairs of oppositely pulsing "satellite" wave-forms, but they are all impossibly on the same "orbital energy-state shell" (See Figure 15, pg.). An interesting thing about these spheric "matter-wave" forms is that the inner central "emptiness" and the outward "emptiness" are both the *unmanifest Energy Wave Field* from which and into which all seondary manifest waves pulse. The spheric "matter waves" are direct manifestations of the Energy Wave Field. "Matter" does not "have" Energy; matter literally *is* Energy that is functioning in a pattern as form *(*somewhat like the form of a whirlpool that is not different from the water's motion of which it consists). Since these are 3-D *spheric* matter wave-forms (presented graphically as 2-Dimensional See Figure 8, pg.), it would be easy for these to be mistaken as 3-D "particles". Although the mathematics might initially be the same, I suspect that with more wave-research in computer simulations, new insights would lead to amplifications and corrections in the math and existent equations. *Schrödinger* was correct that *waves are actual*, and not merely "waves of probability" as *Max Born* maintained. That statistic is derived from wave *intensities* as applicable only to *interference densities but not to interference frequencies.* Light waves from sources at great distances would exhibit a *cumulative red-shift effect of the increasing number of interferences with their decreasing frequencies and longer wave-lengths, along with the cumulative effect of innumerable gravitational fields,* but not as the "recession-expansion" Doppler effect. Since most if not all radiation is emitted outwardly from stars in spheric "shells" of plus-crests and minus-troughs, matter-waves and radiation-waves are both spheric, and *both* are the manifest pulsing action of the *Energy Wave Field* itself.

Einstein posited that "gravity" was the geometry and architecture of space-time. But he also said that gravity was the "warping of the 'fabric' of space-time by the presence of matter-mass. But if all *Energy is equivalent to mass, then the initial* "warping of space-time" does not require the presence of matter. In the EWF, an initial curvature of space-time is initiated by and as the EWF itself, and *instead of curvatures initially appearing where matter is*, matter only appears as an effect of interference curvature densities as in Max Born's "probability rule". *This reverses the initial cause-effect of Einstein's General Relativity.* The waves are actual, *but the higher density of wave interferences effected by gravitational curvature increases the probability of finding spheric matter-waves* ("quasi-particles") at those density locations.

Given that all Energy is equivalent to mass and that mass is associated with gravity, the Energy Wave Field is here posited to be itself the *initial gravitational curvature prior to the presence of matter*,[69] Energy waves that self-interfere and yield spheric matter-waves would add their effects to further warp, twist and contort the gravitational geometry of space-time. Since spheroidal forms of various masses all seem to spin to create a twisting "frame-dragging" effect, the presence of spinning matter will also have its own further effect on the warping of space-time (See Figure 34-C and D). "Gravity" as equivalent to "acceleration" is the first manifest effect of the Möbius-like wave accelerations of the Energy Wave Field, and the *densities of matter* contribute secondary effects to the "architecture" of the EWF Universe. As stated earlier, *if reflection, refraction and spin are all "acceleration" phenomena of Light interacting with matter,* these phenomena must produce locally variable gravitational curvatures. The origin of *spin* however, has yet to be answered. The hypothesis here is that the spin of material bodies and spheric wave-forms (as "quasi-particles") might be imparted directly from a *Möbius-like twisting and curling Field of Energy Waves*, and this would necessarily precede any interferences that might last for varying durations as spheric radiation and matter wave-forms. The self-referencing of the Energy Wave Field at all scales and modes of existence is self-reflected or "bent back" to self-interfere as though two (ie. as "other to itself"), and is iterated at all scales as 1) lower harmonic multiples of an initial higher frequency; 2) a turning "inside-out/outside-in", ("*ever-*

sion/inversion"); and 3) as "the whole in every part" (as in a holograph and in the *Mandlebrot* set.) This "self-referencing" is not only manifest in material corporeal form, but in incorporeal conscious mind. Unfortunately, this "self-referencing" is mistaken as thought identifies a "self" with memory's residue of past experiences. Otherwise fractal-like, the interferences of waves in all patterns and forms always refer back to the *Unmanifest Energy Field*, as well as revert back or "recycle" to interfere again in various new configurations. As space-time is measurable relative only to its contents, everything that exists is inside the Universe. Even the term "Isolated System" implies an outside that the inside is isolated from; but *by definition, there is no outside to Universe.* The *Unmanifest Energy Wave field* as spatially and temporally immeasurable cannot be exactly the same as "the universe" of relative and measurable space and time—or can it?" "Uni-verse" means "one turning"; (the Latin root is *vertere* means "to turn".) Was there ever an actual "beginning" to the turning? If all that is in the Universe at any moment is manifest in wave interferences as "content", then the *Universe as unmanifest Energy conserved is timeless and without beginning or end*, but only the *patterns and forms as effects* which permit the measure of space and time *are subject to change and destruction.* Only the wave interferences as *phenomenal content* can be subject to time as change and entropy as increasing "disorder", however estimated. "Entropy" is applicable only to the interference *form-content* of a timeless vibratory *Energy Wave Field Universe.* Ultimately, there is no absolute distinction between form as measure and immeasurable formlessness.

The Recycling Universe
Nothing seems to be lost from our universe except specific patterns, forms and configurations. The Energy Waves released from old interference patterns are reconfigured as new interference patterns and forms. The Energy Wave Field is pure action, perpetually manifesting as new interference configurations. For example, every element transmuted within stars finds its way into "vacuum space" to combine via radiation into complex molecular compound interferences, some of which are necessary constituents for the DNA "mega-interference molecule". Even in black-hole singularities where form and pattern are crushed to oblivion by the extreme gravity, the energy of those forms released from their configured forms is recycled back to that universal well of the unmanifest Energy Wave Field as mass-

density gravitation. The idea of applying entropy to the Universe itself overlooks this perpetual and eternal recycling that is the universe, evolving by inverting and everting in a manifold of convolutions as though turning itself inside-out. Even "waste heat" lost by friction as well as other non-conserved forms of energy need not become an entropic "thermal equilibrium" where Energy (barring the question-begging) no longer "has the capacity to do work". "Heat," understood as the "flow of energy" as temperature change is only one form of Energy, and as shown earlier, the stasis of "thermal equilibrium" leads to the contradiction that "energy" is both present and not present.

There appear to be only two alternatives for a recycling Universe: *In the first alternative*, after a Big Bang that sets things spinning while expanding either "chaotically" or "orderly", the universe evolves to the point shown in Figure 34-D when space-time contracts as spinning black holes coalesce to one singularity of immense mass density or by some other trigger, followed by another Big Bang and so on. The idea of *entropy* carrying over from one Universe cycle to the next overlooks the *timelessness of Energy conserved* in this perpetual recycling. As each new cyclic Universe goes through its evolutionary changes, the unanswerable question arises as to *whether or not there is some kind of evolution from each cycle to the next*.

An "unstable universe" does not necessarily mean one that is "either expanding or contracting", as has been interpreted from Einstein's equations. What equations *indicate* mathematically depends on the assumed parameters and values that are plugged in, and the *interpretations that can also be varied* as exemplified by the several different models derived from Einstein's equations. Could "unstable" as contracting or expanding also be interpreted as "dynamic", ie. as conracting and expanding at the same time? Is it possible that a universe can be both expanding and contracting as though turning inside-out, somewhat like dough kneaded by a baker? What if there was no beginning or end to a convoluting Universe as in the *second recycling alternative* just mentioned? Instead of *recycling cycles*, the conserved and timeless Energy has the Universe perpetually turning inside-out and outside-in, *everting and inverting* into manifold convolutions in a *"dynamic equllibrium"* as Energy wave interferences configure and reconfigure, creating patterns and forms that are subject to *entropy as change, deterioration and cessation*, only to be replaced by other new interference patterns. All energies released into the overall *Energy Wave field* from "dissolved" form-systems would become as newly formed self-interference patterns and systems. Such a universe would be said to be in a state of entropic "*dynamic equilibrium.*" *The second law of thermodynamics states that entropy can never decrease;*[70] but "waste heat", for example, is just one "form" of Energy. Why would those energy-waves not interfere as *useful input* with other wave systems? In this sense, *entropy does not actually decrease; rather it is "nullified" at the cessation of the old "high entropy" form.* In the ever-changing Universe, entropy neither increases nor decreases overall, *but remains constant, which does not violate the second law.* I admit that I tend to favor this second version, but I realize that spinning black holes might twist and warp space-time to the point where local contractions coalesce to singularity as in the first alternative. It is impossible to tell which—if either—of these scenarios might be correct, but in time, there might be observational measurements possible that would indicate either or neither of these alternatives. Also, in either case, the spheric wave-forms and "macro-cosmic" bodies that spin would still be the effect of a sort of "churning" action of Energy Waves whose twistings initiate a space-time curvature. The American *Charles W. Misner* thought that right after the Big bang, the universe acted "chaotically", so his model was nicknamed the "mixmaster universe".[71] Later, the Russians *Belinski-Khalatnikov-Lifshitz* also offered a somewhat similar idea called the "BKL universe"[72], but these were both more turbulent as effects of a spinning-expanding Universe from a BB beginning (unlike the one posited here). The Universe as an everting-inverting manifold in everlasting dynamic equilibrium is not the same thing, there being neither a "big Bang" nor over-all expansion, nor is there any violent turbulence unless it be in *very slow* imperceptible movements over extremely long eons of time. But I see no way to test this.

Although there is a kind of "energy exchange" between and among systems' configurations, there is no over-all "flow" of energy; it is just that local energy from vanished systems is "released" into the whole *Energy Wave Field*, which is just the disappearance of that particular interference moiré pattern. The self-acting action of this all pervasive Energy Field of Waves is continually and eternally self-interfering to effect new forms and patterns that are subject to the dissolution of entropy. Patterns and forms come into existence and cease to exist as the relative measurability of the space-time universe.

On our own planet life and death is a constant turning inside out, an inversion and eversion, eating and excreting; sex, with "*outer* male" and "*inner* female", *inner* sperm expressed *outward* and *into* the uterus to penetrate *inside* an egg; the blastula forming by *involutions* for *inner* notocord and organs; embryo-fetus-birth as *inside-out*; new cells from *inside*, sloughing off old cells *outside*; even *"internal"* thought *ex-pressed "outwardly"*, seeds sprouting and bursting from inside the earth to the outside of air and sunlight; blossoms opening *inside-out*, and *closing* again, the sun's *outward* radiation from energy *deep inside*, effecting plant photosynthesis, volcanic eruptions, earthquakes, weather; stars *compressing* and generating radiation as heat and Light, *exploding* to spew *its interior* elements *outward* into space where radiation transforms elements into molecules; *throughout the whole universe, all life lives on death in a perpetual singular turning inside out*. It is not unreasonable to posit an eternal Universe that is not merely rotating but turning inside-out to recycle all conserved Energy as self-interference patterns manifesting as phenomenal "things, stuff and processes"; always the free play of Energy in one perpetual and eternal movement that manifests as life and death itself, as creation and destruction in an eternal dynamic equilibrium without beginning or end.[73]

Space

It is a point in Quantum Field Theory that there is no empty space between particles, and that the entire universe can be in a "vacuum state" *where there are no particles at all.*[74] In the *Energy Wave field* model, a universe without any manifest wave interference moiré configuration effects is *equivalent* to a vacuum state of a universe empty of "particles". In both Quantum cosmology and the Energy Wave Feld, there is no "absolute emptiness" of space. The "quantum vaccum state", devoid of manifest particles, is nevertheless a hypothetical state of lowest unmeasurable and unmanifest "ground state" of Energy. This is equivalent to the unmanifest Energy Wave Field with two exceptions: 1) Energy Waves are substituted for "particles", and 2) instead of a "lowest energy ground state", in the EWF the "vacuum state" is predicted to be an equally unmeasurable but extremely *high energy state because of the subsequent decrease in frequencies with corresponding increased wavelengths with each successive interference pattern.* In both, the "vacuum state" might be an "effervescent sea" of short-lived evanescent interferences, as well as no interferences at all. There is an *uninvestigated* possibility that there might be such angular interferences where secondary wavelengths are *shorter* with *corresponding higher frequencies* than the initial wave sets.

It can be asked "Is there such a thing as "space" if there can be no measure?" In Einstein's Relativity theories, space and time as measured are inextricably united; but that "relativity" is dependent on the absolute constant speed of light in all *variied frames of reference, and to relative motions* of *observers' observations* of things and systems (Keep in mind that as atoms, molecules and cells, the observer as well as the observed are both also very complicated complexes of Energy Wave interference patterns). In the unmanifest EWF, there are no "things or systems"; there is no "content" at all except the vibrational Field of Energy Waves. Space empty of content is just the Energy Wave Feld itself. *Isaac Newton* visualized an "absolute space empty of all things"; but you cannot possibly *in your own mind* negate the observing mind that imagines the "emptiness of space" without content; so besides just "Energy", *mind too is ever-present as self-acting Energy (ie. as attention, intention and memory-imagination and thought).* The timelessness of an unmanifest and measureless EWF is not in contradiction with a universe of measurable space-time as size and distance *intervals of objects or time as the measure of change* between positions and differing states of "thing-systems". Magnitude and change belong to a phenomenal world of relativity, but a state of EWF "emptiness" is the absolute fullness of unmanifest Energy as the "plenum" foundation for all manifest phenomena as "reality".

The *relativity* of space-time is a *consequence* of Einstein's posited "*absolute speed limit of light*", regardless of variable "frames of reference". The observer's measure of the observed is relative to the "absolute" speed of Light. But there can only be one absolutely Absolute. The "absolute speed of light" is merely a way of saying that the speed of light is an "unsurpassable limit". Because Energy *is the whole unmanifest as well as the manifest Universe*, it is therefore every "thing"; every "where" and every "when"; thus I've stated that Energy is *omnipresent* and *omnipotent*. However, The Energy Wave Field may be "eternal", "omnipresent and omnipotent", but as we will see, it is only a sort of *hybrid* relative absolute and not The Absolute. But as immeasurable, Energy also "*overflows*" all the manifest *Universe*. As the "crucible of creation", the self-interferences of Energy Waves within their Field effect the initial interference moiré patterns that, fractal-like, continue producing self-referencing configurations at all scales. Since the highest frequencies we are aware of are the so-called "*hard gamma rays*" at the extreme far right end of the known EM spectrum, the Energy Field and its Waves appear to vibrate at an extraordinarily and unimaginably high frequency to be so repeatedly down-shifted in interferences—*unless* time itself changes with each subsequent interference. To "exist" (from the Latin *sistare* = 'to stand") means to "stand out" from a back-ground, where a "*backround*" might be manifest but relatively undifferentiated, or might be an unmanifest background like the *Energy Wave Field* (or the vacuum state empty of "particles"). But "to exist" also means *to be related to other existents as well as to that background itself. All existence is relative*, but the totality of all relativity must be itself "relative" to some Absolute.

Time

Thought as words, like matter and form, is bound to time. The human mind cannot fathom "forever" and "eternity", as it cannot fathom "emptiness" and "nothingness", or it's own ending. But if Energy conserved is indestructible, that *necessarily* means without beginning or end. Is the idea of a temporal universe with a "beginning" possibly derived from the Judeo-Christian *Genesis*? Even the "runaway expansion forever" model has a "Big-Bang" beginning and an "ending" in cold isolated oblivion (or as a recycled "Big Bang" universe). But even the BB can only be a "Bang" of pre-existent Energy, a "beginning" with an extrapolated unending expansion. Both are assumptions from what may very well be erroneous interpretations of the observed red-shift. *So too, the final model in this paper as assumption is no exception; but the wave interference moirés as spheric waves—not particles—is solid and based on observed fact* (albeit a graphic model); it is not a model for the entire Universe, but of Energy Waves in self-interference densities where spheric matter-wave forms are most likely to be found. That is similar to Born's "intensity-as-probability rule", except "intensity" here is a measure of *interference density, and not amplitude squared*.

Time is the measure of change. A universe empty of content can display no change and is therefore without time. It might be said that a "pre-beginning" was the Universe in *potentia*, and that the *interferences* of Energy Waves marks "a beginning", after which the Universe is filled with wave interference manifestations. But this is still thinking in time, and is it possible to not think in time? *Thought is not only the measure of time; thought, as it flows from memory is not only "bound to time"; thought is time.* Time as the measure of change exists only as *consciousness of a present "now" is compared to a past "then"*; but it is still not understood how the brain "records and holds" the past in the present, especially the moment-to-moment "spacious present". Yet what is memory but fragments of an error-ridden chronology of *mental images* of a past? Like history, even one's individual historical past is fragmented like a jig-saw puzzle with most pieces missing, and is it not the same with all "knowledge"? The one sure thing we know about time is that things wear out and break, and one grows older and closer to death from the moment of one's birth; but "Energy conserved" is timeless. *Only interference patterns—including thought and in-formation—are subject to entropy* as the "old age" of time, and time to die.[75]

It is not so much that Energy is invariant in time and momentum (p) is invariant in space; both Energy and momentum are "invariant" in both space and time. The separation of "Energy" and "momentum" is due to the 3-dimensions of space and 1-dimensional direction of the "arrow of time" But regarding special relativity and the *elasticity of time as simultaneity and order of succession*, time as uni-directional cannot be absolutely correct. *Time as unidirectional is perceived strictly within each frame of reference and would certainly be considered to be one-dimensional*; but among innumerable moving frames of reference in the Universe, *the relativity of "simultaneity" and "order of succession" means that variable time is therefore sort of "spread throughout" 3-D space, and in some sense "before and after" are reversible" relative to some of the different reference frame accelerations.* What this seems to do to causality is perplexing.

Entropy (from the Latin *en*= "*in*" + *trope* = *transfomatio*n) is "in transit", a turning" or "change"; *entropy is just a measure of a system's capacity to undergo spontaneous change, It is only in modern physics that that entropy as change is linked with a temporal direction toward greater "disorder"* (however that is measured*)*. The ancient Greeks saw the world as chaos from which intelligence *(nous and logos*) brought order as cosmos. Modern physics has "order" as a low entropy beginning moving forward in time toward higher entropic "disorder and chaos". If I understand correctly (and it is difficult to understand the various expositions of *Ludwig Boltzmann's* definitions of a low and high entropy states as each having certain "degrees of freedom") for molecular unit-particles to be able to assemble in more various ways seems to conform to a low entropy "Big Bang" that evolves toward a state of higher entropy at *higher states of organization* as applied to the entire universe (as described on page 27). According to what I understand, this means that more complex systems having evolved with a greater capacity to explicitly manifest life and intelligence and with a higher degree of freedom as "self-determination", are viewed as higher states of disorder. Despite Boltzmann's hypothesis, although entropy *might* be linked with the "arrow of time", *the aim of that arrow has yet to be determined.* If survival is "programmed" into the DNA of every living cell, might even the once formed self-replicating DNA itself be "timeless"? But what is "once formed" can also be "unformed" and destroyed, and must therefore be time-bound and subject to entropy. In outer space, the radiations would break all bonds of DNA. DNA is a complex of wave-interferences—as are all atoms, molecular structures and cells—and therefore *subject to change and destruction*. It is perhaps an irony that the Universe produces the chemicals for life, but once life forms, the Universe appears to be *deadly* to life-forms.

The Observer

The *observer* is without question the major factor common to both relativity and quantum theories. In relativity, the measure of space-time is dependent on observers' relative positions and motions; in quantum theory, the "outcome" of an experiment is dependent on the *macro*-experimental test design and apparatus, as well as on the macro-observer's intrusive measurement of an invisible "*micro* quantum-world". As mentioned earlier, according to Bohr, the only knowledge of a quantum *micro-world* are the *macro*-world clicks and numbers on an indicator as observed and measured by a macro observer (ie. by a conscious entity). The rest is inference and interpretation from mathematic equations and even these are self-referential; so in both relativity and quantum physics, what is observed is dependent upon the observer.

"*Consciousness*" is defined as *sensitivity to—or awareness of—one's organismic and mental* "self and surroundings", which would be essential to survival; by that definition, even an amoeba or a plant, is "conscious", whether motile or not. *A self is intrinsic to organismic survival.* We do not know how an amoeba for example, is "aware of or feels" a thermal or chemical environment that is "hostile" to its individual survival, but its DNA makes sure it does, and how is that DNA "programmed" to do that? In themselves, sensations are neither pleasant nor painful. In *themselves*, there are no "*sensations*"; these belong only to life-forms programmed for survival with an "impulse feedback awareness" as pleasant or unpleasant "sensations". The question is, with all these electro-chemical signals going through a cell's cytoplasm and nuclear DNA, *how does that organism become aware of sensations of pleasure and pain?* And how does the DNA "remember" in the present a sensation that is now in the past? In a timeless energy-universe, there would be no beginning or ending; but *change as an evolutionary direction* of life-forms cannot be necessarily ruled out; *as direction implies aim and intention*, that intention would manifest in time-bound evolution of life-forms. In this sense, there would be in that intention as well as mere survival itself, a kind of *"universal memory"*—not as a specific "content" as fragments of some particular experiential past in an individual life—but "memory" as a *capacity for a past to be "recognized" in a present as threat or benefit to survival*; memory as a kind of "timeless span" of time" where past and present meet in what has been called a "spacious present". Memory that is "timeless" that is manifest in time-bound life-forms, not only as "survival memory", but also as a potential or capacity for an on-going depth of *self-reflective awareness*, a "bending-back" as a *"self-referencing" intention to be realized*, the *implicit made explicit* and the *universal perceived in—and ultimately as the particular* "endings" within an individual's evolution that are an open door to new beginnings in a timeless recycling Universe at dynamic equilibrium; time as a kind of *Generalized Universal Memory* that manifests throughout the Universe as a capacity for whatever time-bound life-forms that arise where and whenever. *How could there ever be any survival-learning otherwise?* Without some kind of Central Nervous-System (CNS), *how else could there be a manifest intelligence that recognizes Intelligence? Without memory, time as the measure of change* is impossible. Without memory, there could be no awareness of the *movement* from youth to old age. Time is the measure of change, and measure is comparison and without memory there can be no comparison of a past to a present; without memory as time, there is only "is-ing"—no, that's incorrect implying as it does an active and changing present; without memory, there would only be *is—the realization of each moment as a whole unto itself* with no cause-effect connections in a changeless static universe without any such thing as a "dynamic equilibrium"—without even "survival".

As pointed out on pages 20, numbers are the value-measures plugged into an equation, and from the arithmetic and algebraic rules applied, numbers are the result; in this case too the observer is the observed, the measurer is the measured. *But who is this "observer-measurer" upon whom our notion of reality depends? The Central Nervous System*—spinal cord and brain of humans *has evolved from the Universe itself;* in part as a system for survival on this planet Earth, as well as evolving for life elsewhere in this Universe. Considering the probability of intelligent life elsewhere, the likelihood of a kind of "generalized universal" CNS that evolves specifically for survival under a wide range of conditions is not an unreasonable supposition. The transmutation of elements in stars from Hydrogen,

Helium, Carbon, Oxygen, Nitrogen etc. seems to lend credence to the Carbon-base assumption, and it is quite probable that the DNA double helix would be found in all life forms wherever found, the goal being survival to reproduce for survival to reproduce etc. Looking at the evolution of life on just this planet, as life-forms disappear and are replaced by other life-forms, the CNS—especially the brain—has become more complex, more enfolded with surface convolutions contributing to feedback that renders humans capable of hindsight, foresight and planning, with an amazing aptitude for a "how to" survival technology (even to the point where today it seems to have become maladaptive). But the point is that this increasingly complex *instrument of Intelligence* (the CNS-brain) must be manifesting in varying degrees throughout the entire Universe, evolving and developing to a point where *self-reflection as a conscious "bending back" into and onto itself as self-referential recognition of itself qua manifest Intelligence.*

We confront the unknown with a body of acquired experience and knowledge (the known) from which background we translate the *unknown* into what is familiar, making judgements from old "standards" from the past. A "fact" as accessed through our limited senses and our recording and measuring devices is already a "bias" in that the "reality" we humans experience is just a very small and limited portion of the EM spectrum, and very different not only in scale, but also in sensory frequency range from the many various life-forms on our own planet. Our surroundings are experienced quite differently from most other life-forms, such as a bat, a bee, a snake, octopus or amoeba, and although conceived by us as "the same world", it is distinctly different as experienced by each of those life-forms. There are "many worlds" indeed, but many are right here on our own planet Earth. For every scale, for every arrangement and sensitivity-range of sensory apparatus, there is a multiplicity of very different worlds inhabited by lifeforms different from our human world. Even with the technological "extensions" of our senses, what is retrieved is still *transposed* into our own audio and visual range of sensitivities as photos, sounds, screen images, shadows and color patterns. *But one thing all have in common is the instinct for survival on this and on any other planetary systems in the Universe, and that is a manifestation of intelligence.*

In the East Indian *Advaita Vedanta* philosophy, there was an ancient "heretical" sect advocating a doctrine called *dṛṣṭisṛṣṭivāda*, meaning "perception is creation"[76] I cannot say if that meant an absolute creation *ex nihllo*, or the "creation" of a world conditioned by our perceptual apparatus, but if the latter, I see no reason why that should be regarded as "heretical". Light from direct emissions, or as indirectly reflected and refracted by "outside matter", enters the eye to strike the *limited-range* retina where it is converted to electro-chemical impulses that are transmitted to the occipital lobe of the brain where those impulses are somehow "reconstructed" inside the brain to "reconstitue" a visual image that *magically appears out there, with the illusion that the observing conscious-subject* "in here" is *different in essence* from what is observed "out there"; so arises a consciousness *conditioned to division* as "inside-outside", as well as the division of "I" and "Other". (ie. "not I"; it is significant that in newborn infants, the eyes and brain do not "process" 3-D space and time; that processing is *learned* from body-sense experience).

The holographic technology described in the following is an excellent model for visual perception. In **Figure 36-A** (next page), a LASER beam is split by a mirror, one beam sent to an object that is refracted and diffracted as a *data-beam* which is then directed by a mirror—along with a *reference beam direct from the LASER*—to a holographic 2-D film that shows a "meaningless" moiré pattern of linear and spheric waves. By projecting a same frequency LASER "reference beam" through the moiré imprint on film (**Figure 36 B**), a "virtual" 3-D image of the object "magically appears out there" as though in 3-D space; but it is strictly visual and not at all tangible to the other senses. It is as close as we've come to a working analog-model for our visual sense. In this analogy, the "data beam" is obviously the Light striking the retina from reflecting objects; but what would be the analog for a human "reference beam"? The brain itself? Although the physical senses can lead to erroneous conclusions about the nature of reality, the brain only "knows" what chemo-electric signals it receives, or as images remembered from past experiences. How is the brain's internally processed image projected out into 3-D space and time? Is it simply because it *comes from* "out there"? Yet in conscious visualization techniques, *the brain and CNS cannot distinguish the internally imagined images from actual stimuli received from out there.*

Audio and visual perception hears and sees different things as "embraced together in space"; but when attention focuses and things are *named*, we begin to see things as not only different, but as separate and divided, even when re-grouped by similarities; but these groupings too are separate and divided by the concept name; the very *function* of conscious thought is to discern singular similarities and differences, to divide "this" from "that" by what-ever criteria; thought's very function—based as it is on the dualism of language and memory identification—is to separate and divide. In point of fact, *the dividing boundary between "me" and the "not me" is the very same boundary in my own consciousness*. The "not me" is just as much "me"! Both are the very same field of the structure and content of consciousness which itself has become as though divided. *That structure is the very root and origin of all our human divisiveness, and from every "interior" human consciousness it is projected "out there" into our world*, dividing human beings one from another along innumerable "fault-lines". All *bodies* are objects to other bodies. Many cultures differ on how *close* to the body is that "shell" of personal *interior* space that is the "interior" I / me / mine subject. But even the "inside" organs of the body are just as much "objects" as the body's "outside*", and that "outer shell as boundary of the 'me' from the 'not-me' is *inside* each of our own conscious minds. And is it not the case that our very own interior thoughts are themselves *objects* to our own awareness?

Figure 36 A Holography with beam split by mirror into reference beam and to film and diffracted data beam to film

Figure 36 B Same frequency beam through film "reconstitutes" a 3-Dimensional virtual-image of the object.

If our consciousness is divided, and our thought is as much an object for consciousness as the "exterior" environment, then just *where* is the center of "my being"? And then what is "reality"? Is "the world" as *illusory* as *Buddhists* believe? The word "Real" comes from the Latin *res* meaning "thing", so "reality" is the world of things, of objects, ie. the world of space-time *physics*, and our interior thoughts and images are also objects of, by and for the mind". Every act, every process, is a "*turn-*

ing inside-out"; the *inside observer* is the *outside observed. The observer is the observed appears to be true in all cases*; "inside" and "outside" are completely reversible, or rather there is no inside / outside. Here is a paradox from someone anonymous: *"The world is divided into two kinds of people; those who say 'there are two kinds of people; and those who don't."* If the outside "objective" world does not lead to the inside "subjective" of one's owns conscious mind, there can be neither understanding of the world, nor of one's own mind.

The Intelligible Universe[77]

In many disciplines, there are discoveries that indicate an intelligence, whether of man or beast. Anyone working intimately with animals has found evidence of intelligence in those they study. In archaeology, a new-found fragment with inscriptions—even when still undeciphered— is instantly recognized as a having been made by some intelligence, most likely human beings from another time. From apes to elephants, from whales and porpoises, even to invertebrates, insects, and microbes, intelligence is attributed by those who study and work closely with them. All through the animal and even the plant kingdoms, intelligence is discernible in almost all if not all life forms, even in self-replicating DNA. We humans who attribute such intelligence to ourselves, still do not understand the mysterious long sequences of DNA that seem to have no current function. Some scientists most likely have a personal belief in something beyond the physical world; but as "scientists", most if not all share an *implicit* assumption (and faith) that the manifest universe has a comprehensible *intelligible order*, and that scientists can learn to understand that order; otherwise why would any of the sciences even exist? Anything "intelligible" has always and everywhere indicated "intelligence", and only an intelligent mind can discern intelligible order as intelligent mind. *Then how is it that the intelligent mind of humans—and scientists in particular— cannot discern and admit that same intelligence in all nature, and certainly in the universe from which our DNA and brain are derived?*

There is always a metaphysical assumption underlying every belief. When the first law of thermodynamics was formed, the concept of Energy moved physics more deeply into metaphysics. Energy, being invisible, intangible and immaterial, is an unknown that is inferred from its many manifest actions such as mass, work, force, radiation and matter, and as extended by etymological analysis, is here considered to be *self-acting action without* any actor and only as action itself. As the capacity to do work, *Energy is unmanifest*, and is known and measured only by its manifest work. *Entropy as the capacity or measure of a system to undergo spontaneous change*, does not in itself imply any specific direction, *nor is there a definite correlation of entropy with time as direction, only as change.* The second law of thermodynamics only *states the probability of entropy increasing with time*; but entropy is not necessarily the "causal foundation", nor even the "arrow of time" as direction. *Ludwig Boltzmann's* "degrees of freedom" for molecules' "assemblages" does little to clarify how "order" and "disorder" are to be interpreted.* The Greek universe began with *chaos*, and manifest intelligence (ie. *nous*, as mind) made *cosmos as order from the disorder of chaos.* Only in modern physics is entropy as "spontaneous change" considered to be change from cosmic "order" toward greater chaotic "disorder". Boltzmann defined "order"(if I understand correctly) as having a "*higher probability for greater degrees of freedom*" (ie. as *low entropy), with more probabilities for molecular arrangements, and higher entropy "disorder" is regarded as "fewer probabilities for degrees of freedom" of arrangements.* This of course conforms to *a low entropy Big Bang*, with an inevitable "direction" toward greater disorder. According to that model, gravity's clumping and "structural complexity of arrangement" has a lower degree of freedom (ie. *higher entropy*) than a more *probable random* arrangement as a *lower entropy.* Would a Universe from which the chemistry of life, and a CNS structure that manifests intelligence that can look back at the source from which it evolved be toward greater disorder? Energy as conserved has no limit, and as the formless source of form, Energy cannot be measured or estimated except and *only* in particular manifest transactions between systems where "local" energy is assigned "plus" and "minus" which must always balance out.

Either the universe is disordered and it is *human* intelligence that creates or imposes *order out of chaos*; or the universe is intelligible order and *humans as mere observers only discover and catalogue the order that is independent of the observer*; or "intellig-*ible*" implies an "intelli-*gent*" order, and therefore an *intelligent universe where the observer is the observed.* Scientists are generally not wont to ascribe intelligence to nature and the Universe, *but is not the intelligence of humans* (out on a limb here!) *derived from a Universe that gives birth to life itself?* Yet as outlined above in "*the recycling universe*", amino acids form in interstellar and galactic space out of the "excretions" of elements from exploding stars, and the very first elements to form within stars from Hydrogen are Helium, Carbon, Oxygen and Nitrogen, Four (counting Hydrogen) of the basic elements for DNA, proteins, and life. If the universe be intelligible at all, then the Universe must also be intelligent or have intelligence in it. According to the dictionary, intelligence is synonymous with "*Incorporeal Mind*". Is it not this intelligence of the universe that has evolved an immense space full of galactic clusters with some stars as suns for solar-systems around which some orbiting planets and satellites have conditions conducive to life-forms evolving in a direction of greater complexity, where a "survival CNS" *also* has a feedback capacity for mind to reflect on the origins of life, the brain, of the Universe and of Mind"? The question then becomes "Who (or what) is reflecting on whom?" If Energy is self-acting action, Intelligent Mind is self-referent intention. We do not know the depths of our own mind, but Mind does; therefore Mind is the "*omniscient* knowing" of the darkest secrets of the human mind as well as the secrets of the Universal Cosmic Order. As life-forms adapt to the changing conditions through time, many become extinct; but intelligence keeps manifesting at ever higher "degrees" to points where a brain capable of reflective feedback has the capacity as an instrument of Intelligence to perceive "mind" as that same self-referencing Mind of Universal Intelligence. *The Energy Wave Field* as a model for the unitary ground of all phenomenal existence, should now be referred to as *The Mind-Energy Wave Field. Mind's Energy as attention and intention is the self-actualizing of Mind's intent.* The very impulse to wonder, to ponder and question refers back to the questioner as that very self-referencing reflection of mind as Mind. It is not possible within reason to use intelligent mind to negate or refute Intelligent Mind", nor is it possible since the brain has essentially evolved from the Universe through DNA to *not* ascribe intelligence to or in the Universe itself. The *brain is not the origin of intelligence; the brain is an instrument of intelligence, and the fundamental human problem is a problem of "mistaken identity."*

What if there is a direction to evolution *besides Charles Darwin's* mere "adaptation to environment"? (not that that is insignificant; but it is not the most significant part of evolution.) What if *teleology* is not directed toward any final "Omega-ending" *(as Fr. Pierre Teilhard de Chardin* proposed), and instead, sentient life throughout the universe evolves or awakens (even within the same species) *at different "rates and times,"* individually reaching a point where a "generic" and universal "CNS" can explicitly manifest that Universal Intelligence?

An entire Genus might perhaps have that capacity, but a conscious wonder leading to intentional reflection as a necessary prerequisite for *actualizing* that capacity, and perhaps to varying degrees, intelligence would manifest as Universal Mind "in the flesh" of some individual(s) at different times. This "Intelligence-as-Mind" *does not imply a "being," or "entity"; the word "God" is especially rife with a legion of interpretations and corporeal images*. In a very real sense, everyone has their own individual god. but within *The Mind-Energy Wave Field, intelligence* as self-referencing Mind and self-interfering Energy Waves *is intrinsic to all energy and matter wave-forms*; between "Spirit-Energy" and "flesh", there is a difference, but there is no *division*. Energy that functions in self interferences effects spheric wave-foms as "Matter-Energy" is formless and timeless; and *entropy* as change can *only* apply to "material forms". Mind is Energy and Energy is Matter, and Mind is *the Limit* beyond which mind cannot go. Whatever might or might not be beyond Mind as some "higher Other", Mind cannot reach "that"—*unless Mind as such is that very "That."*

SUMMARY

If, according to the law of Energy conservation, Energy is indestructible, it must therefore be timeless and eternal. In this paper, Energy is initially hypothesized as the unmanifest foundation of all manifest Universe and all subsequent evolution. The entirety of this paper stems from an earlier observation of graphic wave patterns printed on 2-dimensional transparencies that when overlaid produce secondary moiré effects of which two kinds are the focus of this paper: 1) "rectilinear" or "curvilinear" moiré configurations that are almost always of lower frequency and longer wavelengths than the initial interfering wave sets, which clearly demonstrates that the Compton effect does not support Einstein's "energy particle" assertion, and instead upholds the *wave theory* of energy; 2) of even greater significance are the *oppositely pulsing pairs of spheric wave forms* on either side of the *initial* spheric wave in interference with a rectilinear or curvilinear wavetrain. The opposite directions of the pulsations are dependent on the relative direction(s) of motion of either or both initial wave sets. These in and out pulsings are most probably equivalent to "charge", but the s*pheric* wave-forms so easily mistaken as "particles" might also actually "spin". It was from the observation of these patterns described above that led to the hypothesis of an *Energy Wave Field* and this work.

Many believe that "mind" is a sort of "epiphenomenon" of the brain's processes; the brain itself however, is an *evolutionary result* from the Universe. If the Universe and Nature are not thought to be intelligible, *it follows that that is necessarily the end of all science.* In archeology, when an inscription is discovered, even prior to it being deciphered that inscription is taken as a sign of intelligence attributed to ancient humans. Why then, is an intelligible Universe not taken to signify intelligence? Apparently only because of a contradictory bias in the sciences. Physics appears to still be grounded on the old premise that all *chemical mechanics* can explain the origin of life and intelligence. In this work, intelligence is ascribed to be or be in the Universe just as an ancient inscription is ascribed to some human intelligence. Are we so arrogant and full of hubris that we believe it is our own human intelligence alone that attempts to "read" the Universe? What makes a mind individual as "yours" or "mine" is but the tattered remnants of our particular past experiences whether remembered or not which make up our "self-identity", which is a case of "mistaken identity". The use of the capitalized word "Mind" refers not only to an individual mind, nor only to intellect; the German word *Geist* means both "Spirit" and "Mind" ("spirit" being used here as a near equivalent to "energy"). The Chinese *hsin* is translated as "heart-mind" which includes feeling as well as intellect. As for the Universe, it seems to be much like a hologram, with Light as emitted and reflected from matter as representative of the holographic *data beam*, and Mind-Energy being the *reference beam*, with the film and screen being the Mind's brain and senses. Touch is basically the electro-chemical effect of molecular density waves interacting with molecular density waves. *But significance and meaning cannot be explained by mechanics or electro-chemistry.* What is going on in Universe as we know it is on-going, a "work in progress", perhaps without beginning or end. The Universe and all its subsequent Energy-wave forms are *iterations* of One self-referencing Mind-Energy Wave. As presented in this work, the wave interferences clearly show that *between Energy and matter, the difference is not a division or separation; spheric matter-wave forms are a direct effect of Energy functioning in patterns* that are subject to time and entropy as dissolution. That being so, then Energy as Mind and as matter-body are not the "division" as heretofore posited. The material body is form subject to time; Mind-Energy is formless and timeless; ALL pattern and form is the effect of Energy Waves functioning in and as moiré interference patterns as "information" as well as material-form. The psychological structure of the the "I / me / mine" entity are also subject to time; when the pattern dissipates, that Energy is "released" from its pattern, and is then no different from the eternal and limitless EWF. I had previously referred to spheric waves as "matter-waves", but when EM energy is emitted from spheric stars, it too is emitted in "spheric wave-shells" of crests and troughs, just as with the oppositely pulsing pairs of matter waves; the spheric wave form is therefore common to both matter and to EM energy. At extreme radii from the emission source, the spheric "curvilinear" waves of EM energy would appear almost as rectilinear wave-trains. The question in the Preface is posed again: A supernova or galactic explosion would radiate "shock waves" in all directions, observable only as disturbances of physical bodies. What would be the *difference* between those perturbations being inferred as "Energy waves" or as "Gravity waves"?

The Mind-Energy Wave Field can be modeled in several different ways as a *One that self-interferes as though two* (**See Figure 37 A**) As a spheroidal (spheric or ovoid) wave that twists, bends and curls Möbius-like to initiate *gravitational curvature* as well as *secondary interference moiré waves*, it also imparts *spin* to those spheric and spheroidal wave-forms (**37 B**). Another way is as *an ovoidal or elliptical "cosmic egg"* emitting waves that self-reflect interiorly, effecting a secondary focal-point that also self-reflects back and forth as a *source.* **37 C** is a third model wave-train *focused* by a lens, although this is less satisfactory because it introduces an "outside object" unless the lens is considered to be the "Mind's-eye". For myself, I prefer the first alternative (A), whether as a spheroidal or ovoidal wave-form. Besides the multiple effects of the "Möbius twisting and bendings", from that One "self-contorted" spheroidal Mind-Energy Wave, both a *secondary spheric wave-form and several rectilinear wave-trains are simultaneously generated* as shown in Figure 37 A (The illustration is more of a self-interfering Möbius strip than a spheroid; but imagine all open-ends as closed and that contorted strip becomes a contorted sphere or egg, more-or-less).[78] This singular pulsing wave is equivalent to a timeless Universe that is turning *inside-out and outside in* as an unmanifest *Mind-Energy Wave Field* (equivalent perhaps to a "quantum vacuum state" of a Universe devoid of "particles"; but the *Möbius-like twist and bend in and out* also effects a simultaneous *pulsing in and out* that is iterated in the secondary oppositely pulsing spheric-wave pairs as shown in Figure 35. What is the timeless unmanifest, *self-manifests* as an apparent and illusory "in time duality". "Accumulating all the quantum fluctuations that are 'allowed' by the uncertainty principle results in an *infinite vacuum energy density*:[79] What makes the infinite finite is measure, comparison of something to a standard or ideal. What the wave interferences demonstrate in this work is that the *unmanifest self-acting action of conserved Energy* is the same that folds into self-interference effects of spheric moiré patterns *as matter and as EM energy*. Form and formless are the same—yet distinct; Mind is Energy and Energy is matter and Mind is matter as manifest brain. If the whole Universe of Intelligent Mind is in the brain, then what can "limit" mean? Is it all just our own Mind that waves?

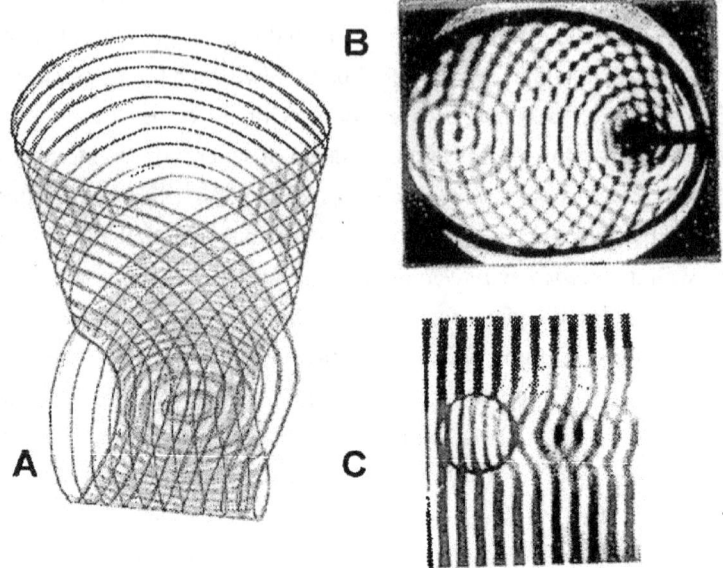

Fig 37: A—Möbius twisted waves with moirés; B—elliptic reflected waves; C—waves with lens

POSTCRIPT

(AN EXTENDED SUMMARY OF MAIN TOPICS)

1—Rules of Mathematics

In 1811, *Karl Friedrich Gauss* wrote "We should never forget that the functions, as all mathematical combinations of concepts, are only our own creations...they are *arbitrary conventions for the purpose of preserving the formalism in the calculus.*" (Waismann, p. 39) One 'notorious' example is the rule of signs as applied to the natural world (the world of the true *physicist*); a 'minus' times a 'minus' yields a 'plus' cannot be proven and fits Gauss's statement perfectly. Apply that to a motion in the opposite direction of a 'plus direction' and in reality you would actually get a greater minus direction; apply that same rule to your bank account, and you'd be pretty far in debt. The *convention* in physics is to graph left-to-right as "plus" and from right to left as "minus". Using this convention, a car traveling 50 miles left-to-right, then returning right to left (or in time upward from and then downward to "zero") would have traveled 0 distance. The rule of signs is applied to gravitation, to velocity, everywhere in physics equations. Try applying that to a minus times a minus temperature and venture out in a swim suit or less. The *point* is in the last part of Gauss's statement, "*...to preserve the formalism in the calculus.*" Is it any wonder that when we make a measure and enter the variable numbers into a formal equation, we get—eureka!—numbers that always seem to come out right? But it is the *conventional system* of mathematical rules that makes it so! The formalism is set up that way. The *wonder* of it all is that men are sent to the moon and probes are sent into space using those mathematical conventions, including the rule of signs. Even Netwon's Laws of motion and gravitational equations seem to work, in spite of his arbitrary and homocentric bias as to how the "useful work of humans" is measured as "positive" *against* nature's "negative" processes!

2—Energy, Mass and Matter

The crux of the entire problem is contained in this statement: "*Furthermore, the properties of matter that we call mass and energy do something extraordinary: They inffluence the vey essence of spacetime.*"[80] (emphasis mine) This statement has it backwards, upside-down and inside-out, as does the whole endeavor of physics-cosmology from it's beginnings. The primacy of 'matter" is incorrect. *Matter* can be created and destroyed and is therefore subject to *time* and *entropy*. *Energy*, being conserved, is not subject to entropy, has no relation at all to entropy, and being also *mass*, is the source of the *initial*, and *more-or-less even* distribution of the curved geometry of spacetime. In the *Energy Wave Field* hypothesis, there is no difference between vibrations as "energy waves" and "gravity waves". These waves tend to interpose in interference configurations more frequently so that "matter"—being merely some of the many forms of energy-wave interferences— is more likely to be found in those more curved spacetime regions. Matter is entirely an *effect of Energy-wave interpositional interferences*, but it is a *direct effect*. The quoted statement can be corrected as "*The properties of Energy that we call "work"* or "action" and "*mass*" influence the initial curvature of spacetime which determines *where* "matter wave-forms" are more likely to be found, and which cumulative interferences and consequent gravitation increases further influence the *variable curvature effects* throughout spacetime."

3—The "Big Bang" Beginning:

The universe is hypothesized as having a beginning in a "Big Bang" when all matter and energy exploded into existence and "time" first began. The question is how can there be an explosion if there is no energy to bring that about? Is there anything more energetic than an explosion? What happened to the "law" of energy conservation that states that "energy can be neither created nor destroyed"? At a "point-seed singularity" of infinite temperature and mass-density, all "laws of physics" break down. ($1/0=$ "infinity"; $1/$"infinity"$= 0$, and "infinity" x $0 =1$.) In the Energy Wave Field (EWF) hypothesis, Energy is self-acting action and energy as *invariant in time* ultimately means that energy is timeless and eternal. ("momentum" as invariant in space is just the *inertia* of energy-mass in motion.) Since the whole universe is just the EWF, there can be no "beginning" except for the *manifest effects* of energy such as matter and radiation, which are subject to both time and entropy (however the latter is described). What has a beginning, also has an ending, and Energy that does not have a beginning cannot end. Matter does not "have" energy; all matter is Energy that is functioning in a more-or-less fixed configuration. It is not "energy" and "matter"; *all matter is Energy and nothing but Energy*.

4—Entropy:

The term comes from the Greek *en* ("in or at") + *trope* ("transformation, turning or change"), and was originally defined in the 1978 1st edition of *The American Heritage College Dictionary (AHCD)* as "*A meaure of the capacity of a system to undego spontaneous change.*" By 1985, the *Concise Oxford Dictionary's* definition read "*A measure of the degradation of the universe; a measure of the unavailability of a system's thermal energy for conversion into mechanical work.*" By 2004 in the 4th edition of the AHCD, entropy was defined as "*a measure of thermal energy unavliable to do work; the inevitable deterioration, disorder, or ranbdomness of a system*" [undeline mine]. The question-begging goes something like this: "In thermal equilibrium, there is no longer the capacity for energy transfer as "heat flow", so nothing can happen anymore; however, this is not a nullification of energy conservation because energy is still present; *it's just not available to do any work*". But the very definition of "energy" in physics is "*the capacity to do work*". If energy is "unavailable" to DO work, and "nothing can happen anymore", *there is no capacity to do work*, so energy cannot "be present", thus violating the conservation law. However, *if there is any temperature at* all above 0 Kelvin, then energy must be present. So we have two major contradictions: 1) the law of energy conservation is violated if energy is "unavailable to do work", and 2) at the same time energy as any temperature above 0 K seems to be present. Energy as both present and not present at the same time is NOT a "quantum superposition", but a logical contradiction. There is yet a third contradiction using the original definition of "entropy": if entropy is spontaneous change and a "high entropy state" of thermal equilibrium *cannot change spontaneously now or in any future do any work, not only is "energy" nullified, but so is "entropy"!*

John Von Neuman is reported to have said in a conversation withn Claude Shannon that "— *nobody knows what entropy really is, so in a debate you will always have the advantage.*" (wiki/History_of_entropy, pg. 18 of 27 pages). Who has the arrogance to say that the universe inevitably "progresses from order to disorder"? Call it "degrees of freedom", but it is still a judgment grounded in homo-centric hubris to state that "natural processes" on this planet and in the universe are "random disorder". "Entropy" initially meant "spontaneous change", which means the same as "evolution" How can there *not be order in whatever the universe does*? The real disorder comes from human beings on this planet, in both thought and in behavior; whatever nature and the universe do can be nothing but perfect order, whether we like it or not. (see also the last paragraph of 4A below.)

5—Gravity:

A—gravity and energy:

What is the difference between "energy waves" and "gravity waves"? When there is a super-nova explosion, or two black holes converging, or two galaxies merging, the waves of disturbance that radiate outward through the spacetime universe are waves of what? Are they not ripples in the geometry of spacetime? But the spacetime universe is all energy and nothing but energy that manifests in different ways as different "stuff", so are those ripples waves of energy, or of gravity? Is there a difference? We say that mass-density effects the architecture of spacetime, which is gravity; but mass-density is not only matter and radiation; energy itself is mass, and it is being discovered that the distribution of mass-density, as energy-wave interference mass and as matter mass is variable throughout the universe. If this is so, then energy-mass density as gravity varies from spacetime region to region. Einstein's "principle of equivalence" states that no observer enclosed in any kind of windowless enclosure can tell the difference between gravity and acceleration. If acceleration is the action of energy as gravitational "force", it is posited here that there is not only equivalence between energy and gravity, but actual identity, separated only by terminology.

If the ultimate gravitation is a black hole, which holds radiation as well as matter in its grip, and which ultimately means the destruction of matter (as spheric waves) and the release of energy (that has been functioning *in* and *as* form) back to formless and free energy, that of course is "entropy", but only as it applies to the destructibility of form that has been created by wave interferences. Black holes seem to be the destructive, energy-freeing "pole" of its opposite, the creation of radiation and form by the self-interfering action of self-acting Energy.

B—gravitational mass-density increase:

If there is no distinguishable difference between Energy waves and Gravity waves, and if the unmanifest EWF itself is mass, then gravitational acceleration would be an immediate manifest effect. But manifest as waves interpositioned in conjugation (ie. "superpositioned"), the secondary interference configurations would be of longer wave-lengths and lower frequencies, *but with twice the mass-density of the unmanifest EWF* (especially as secondary spheric wave-forms). *Every successive wave interference would produce an increase in mass-density (and "weight"), with a concomitant decrease in velocity.* The *initial* curvature of the unmanifest EWF spacetime would be more-or-less spread evenly throughout an unmanifest universe, and in this state there would be either no wave interferences at all, or more likely brief intermittant wave interferences as a roiling effervescent sea of evanescent interferences that is perhaps equivalent to the "quantum vacuum state" of an un-manifest universe. Once secondary conjugate waves of increased mass-density continue to "self"-interfere again and again, and as wave-complexes build up into "heavier" atomic wave-structures (eg. metallic and crystalline lattices), mass density keeps increasing. The twist and bend of the geometery of spacetime becomes more and more regionally variable, so that what we call "gravity" varies from region to region as mass density increases at different rates in different regions of spacetime. The increase of variable curvatures would result in even more conjugate wave interferences, many of them as spherical "matter waves", so according to Max Born's "probability amplitude", there would be a higher probability of finding spheric matter wave-forms in those regions of high interference-density. Where interference densities increase, there mass-density gravitation will continue to increase, forming a network of manifest matter and hollows or voids.

C—gravitational red-shift:

It is currently believed that at the center of all spiral and barred spiral galaxies lurks a massive black hole. This is a gravitational pivot that acts like an "engine" for a galaxy's rotational motion, and is what is meant by "gravity's constraint". Estimating distances by an assumed "standard candle" would yield a Doppler-interpreted red-shift as increasing recession the farther into deep space one looked, and the conclusion would naturally be an explosive "Big Bang" that set everything in expansive motion—and it would be wrong. For the observer outside of deep-space gravitational fields, time in deep-space is slower, and consequently gravitational red-shift is greater. Gravitational red-shift is not just the effect of a few masses; in deep space there are innumerable sources of high mass-density and an incalculable collective of overlapping of gravitational fields effecting huge red-shifts whose variations might correlate with distance, but more significantly with the regionally varying distributions of gravitational mass-density.

D—galactic spin:

A black hole at the center of a spiral galaxy acts as the gravitational pivot for the galactic disc's rotational spin; but the outer periphery seems to rotate faster than Newton's "inverse square law" predicts. Would not the immense gravity of a central black hole also constrain the motion of surrounding bodies in close proximity to that deep well gravitational strength? As viewed from "outside", time appears slower in that gravitational well, and faster farther out, so rotational spin would appear relativistically faster than calculated by Newton's "law". Also, do we not use gravitational bodies to "slingshot" away from that g-well? The constraining force of gravity would be less at the galaxy's outer edges, so combined with the differential time factor, the peripheral edge should be expected to travel faster the inverse square law predicts.

E—"Photon" Spin:

The spheric wave forms effected from the interferences in the EWF most likely spin, since the Energy Field of waves self-interfere by Möbius-like twisting and bending. In all spins there are three possible axiis—one vertical and two horizontal; however with "photons" of light at lightspeed, the direction of spin cannot be in the direction of travel, the reason being that spinning one way would be going backwards at *slower* than lightspeed, but spin the opposite forward direction would have to be *faster* than lightspeed to "catch-up" as light travels forward (so the argument goes; but wait, isn't the photon supposed to be a "particle of light" itself?). The only spin remaining is when the *axis of spin is in line with the direction of propagation*, and the spin itself is not in that direction, but circular *around the axis and line of travel* (This would of course apply the same way to spheric waves of light or energy if they had spin). What this would mean is that instead of EM-energy waves moving vertically up and down or horizontally back and forth, circularly spinning waves would travel in—and be polarized as—a corkscrew spiral—which would also be the initial effect of the Möbius-like twist and bend of the Energy Wave Field's gravitational architecture of the space-time universe.

F—circular polarity and spiraling propagation:

The unmanifest Energy Wave Field is present everywhere, and all things created are a direct effect of wave conjugation that creates a secondary wave-trains or wave-forms (depending on what kind of waves are interposed). Because every spherical body spins and because no one has answered the question of where spin initially comes from, it is conjectured that the wave-action of Energy itself twists and bends "Möbius-like" so as to self-interfere. Circular polarity is thought to be a special kind of polarization; but since light passes through diagonally polarized filters after passing through vertical and horizontal polarizing filters, it is assumed in the EWF that all electro-magnetic (EM) radiant waves are circularly polarized. That would also mean that the trajectory of light waves—and of all Energy waves—would be as a corkscrew spiral, like the constrained polar beams of synchrotronic radiation from a galactic center's black hole.

"Shear" is a distorting twist that light follows as "geodesics" in gravitational fields. In 1962 Ranier Sachs simplified Einstein's equations by omitting the distorting twist-effects of shear. The question is, "Do such "simplifications" give us a valid picture of "nature's reality"? The EWF hypothesis considers the shear-twist (ie. corkscrew-spiral) as *significant* to understand gravity as well as the "self -folding" interactive interferences" that effect spheric matter-waves" (see end-note 37, p.62).

6—Dark matter, Dark energy:

Positing an unmanifest Energy Wave Field is more-or-less equivalent to a "quantum vacuum energy state" of the space-time universe, and most likely neither one nor the other can be actually verified; but positing "dark matter" and "dark energy" to explain a "missing mass" seems to be sheer desperation to explain an assumed increasing rate of expansion of the spacetime universe. In point of fact, "energy" itself is invisible, intangible and immeasurable and is only *inferred from observing the effects of matter* such as motion, radiation, gravity and other manifest phenomena. As the capacity to do work, the energy concept can be extended as "self-acting action" without an actor, a verb without a subject. Etymologically derived from the Greek *en* ("in or at") + *ergon* ("work"), it literally means "in or at work", but there is no one or no thing, no "subject", doing that work. It is difficult to conceive of action without an actor, which makes energy itself rather "dark". Why is it necessary at all to posit a "dark matter" *which does not interact in any known way with radiant energy, has no chemical signature*, and *is only inferred* from observed perturbations of visible matter; or why posit a "dark energy" when *inferred* energy is itself already invisible and "dark"? No one wants to go back to the drawing-board and begin anew, but isn't that how strides if not leaps have always been made in science? Positing an expanding universe from an *assumed conjunction* of two different independent observations amounts to piling assumption on assumption, supposition on supposition, then inventing something else to account for what is likely our own incomplete account of mass-density. First, there is a whole unmanifest field of *Energy-as-mass* that cannot be measured (more-or-less equivalent to a "quantum vacuum state"); second, there is what amounts to a gravitational unknown to account for "anomalies" that are really indications that our knowledge is incomplete. Thirdly, there is the fact that all EM radiation as "Light" is visible and measurable only by Line-of-sight, which makes our estimations of mass densities mere guesswork.

7—Dynamic equilibrium and a recycling universe:

The second law of thermodynamics states that entropy cannot decrease and can only increase with time, or *at best remain constant*. But there are cases where local entropy can and does decrease, albeit ostensibly at some over-all increase. The term "waste heat" is derived from the old steam-engine days, but the "accumulation of waste-heat" in the over-all universe s pure conjecture based on the *re-definition* of "entropy" (see "entropy" above). The EWF posits a universe that is timeless but not static. In the EWF universe, everything is recycled with entropy decreasing and increasing "regionally" but remaining constant over-all. There is sufficient evidence for this in the processes of the currently known universe, and *if energy conserved is truly invariant in time, then entropy cannot be applied to the unmanifest energy universe, but only to the manifest effects of the work-action of energy.*

Energy lost as "waste heat" from a system can be used in or by another system in a different way for a different purpose. Systems come and go, but the EWF universe is *The System of all systems*, however it may change. With mass-density varying throughout the universe, rates of expansion must vary accordingly but also with areas of contraction ie. a universe that is in constant spontaneous change, turning inside-out and outside-in at the same time in different places in a timeless "entropic" state of dynamic equilibrium.

8—Mass at rest and mass in motion (kinetic energy, or Ek):

Mass in motion is "relativistic mass" (M); (m) is the rest-mass, or the inherent or intrsic mass of a body; but "m" is *still the Ek and Ep* of the oscillations of the sub-atomic "internal constituent particles" (ie. protons, neutrons, quarks and electrons) of an atom "at rest".

In the EWF, (m) is the rest mass of a spheric wave-form as a "standing matter-wave" when not in motion, but includes the in/out of Energy-Wave n-pulsations/sec as frequency, and amplitude squared as intensity of the wave. As there are no "Ek or Ep of particles in motion or position" in EWF rest-mass "m", Ek is eliminated from the contradiction inherent in "particle rest-mass". Ek only applies *when a whole spheric wave-form is in motion through space-time*. A question is *"what is the difference between the "Ek" and the "momentum" of a moving mass?"*

9—Interpretation:

I was once expounding on some topic with a friend, and when I had at last finished, she replied "I completely agree with my interpretation of everything you said." Politicians like to say that "Truth is beside the point; everything is perception." Likewise, "fact" is one thing; "interpretation" is another. From Einstein's equations many different kinds of universes are interpreted. The interpretations of the observed cosmic red-shifts yield different kinds of universes. Mathematical rigor in equations is nevertheless *subject to conceptual interpretations*, giving rise to "schools of thought" as in quantum theory. Observation of the effects of light (and all EM radiant energies) and matter divide into basically two interpretations: particle or wave, or the Copenhagen interpretation of quantum theory where both particle and wave are "complementary", each manifestation entirely dependent on the design of the experiment performed. The entire *Energy Wave Field model* is derived from observations of graphic line-patterns that are *interpreted* as equivalent to invisible vibrational energy waves. Belief is the underpinning of interpretation, which then reinforces belief. Scientists, being humans first and foremost, are as subject to individual and collective biases as much as those with religious, economic or political beliefs, or of life's purpose(s). The eye of the beholder is more often subjectively partial than impartially objective when engaged in self-examination, as well as when looking at "external facts". Interpreting each other most often leads to comedy or tragedy, sometimes to both, sometimes to understanding, and maybe sometimes to Truth.

10—The Quantum:

When Max Planck first introduced the idea of an indivisible quantity that occurred when light interacted with matter (the hypothetical "oscillators"), the "quantum" was applied to spin and to electron energy levels in atoms, but never as "particles" until Einstein posited the "particle of light" and then as "particles of energy" itself (In those days, "energy" was just all EM radiation). Hardly any colleagues believed or accepted Einstein's "point-particles" (later named as "photons"). Today, photons are taken for granted, but it took about twenty years and the Einstein interpretation of the Compton experiment for colleagues to finally accept the photon as a point-particle. The quantum had become changed from an indivisible quantity or ratio to a point-particle, but the particles energy-quantities were either whole numbers, 1, 2, 3, 4 etc., or partial-fractions that are familiar to musicians as *ratios of the lengths of strings* stopped at specific intervals such as 1/2, 1/3, 1/4, 1/5 etc, and it seems that these fractional "quantas" are akin to if not the same as the *harmonic overtone multiples of a fundamental wavelength / frequency of vibrational "tone"*. In The Energy Wave Field model, it is maintained that there is a fractal-like iteration of universal proportions, forms, frequencies of vibrations in general that recur at all scales throughout the evolution of the spacetime universe. This is what I anticipate the "quantum" to be, and like a holographic plate and the Mandlebrot "bug", the whole is in every part throughout the manifest universe. That might be kept in mind.

11—Origin of the Large-scale Cosmic Structure

Known galaxies mapped at about 2 billion light years radius reveal a weblike network of matter with bubble-like voids, and with high-density "nodes" that resemble a satellite map of urban centers connected by transportation-communication lines, with low-density "rural-voids" predominating. "Since Energy and mass are equivalent, any initial gravitational effect of the unmanifest EWF would be more or less evenly distributed. But the interpositioning of wave interferences that effect secondary linear wave-trains—or more significantly—*spheric "matter-wave" forms*, create a local but minute difference of mass-density gravitation that would curl spacetime sufficiently to promote more interferences with the secondary wave, further increasing mass-density whose gravitational effect would continue the exponential growth of gravitational mass-density. More effected waves would also cluster into *sub-atomic complexes, and in time would form complete Hydrogen atoms*, followed by the formation of stars, clusters, galaxies etc.. In this way the skein-like filaments and bubble-like void structure of the universe is formed. Probably the same kind of process of protein-formation created the elemental sponges and same-cell "colony" complexes. Max Born's amplitude (or intensity) "probability" is applicable here; where there are more wave-interferences per units of time, there you are more likely to find "matter." The entire "vocabulary of forms" is fractal-like at many different scales, based on universal growth-patterns via universal proportions and ratios.

12—Intelligible Order and Intelligence

In anthropology and archaeology, intentionally chipped or ground tools, symbolic markings, pottery, metallurgy, burials and building sites are the intelligible works of intelligent beings. In the sciences, human intelligence can understand only what has an intelligible order, and in physics and cosmology, and even in biology, nature and the cosmos are regarded as intelligible. If that was not the underlying assumption, none of the sciences would ever exist. But intelligibility always signifies intelligence, both of the observer and of the observed. Recognition of any intelligible order is a sign of intelligence, again of both observer and the observed. So my question is why do physicists and cosmologists, and even some biologists, flatly refuse to ascribe intelligence to nature and the universe, when it is from that very universe that the elements for the body and central nervous system (CNS) of human scientists are derived? We are indeed "born of the stars" and made of that very "stuff of the universe"; how then can the human brain deny that fact? John

Wheeler was partly correct in saying that the universe is looking at us looking back at the universe. But in the EWF, Energy seems to have no specific direction—or does it? Given Darwin's adaptation to environment, there still remains throughout the universe a seemingly inevitable nisus towards the manifestation of greater degrees of intelligence. This is not at all the same as the homocentric "goldilocks anthropic principle"; "intelligence" applies to all life-forms throughout the entire universe. Humans are not singularly special; we are just one of many, but apparently not manifesting the intelligence that we think we are or have, since even our scientists do not recognize that any intelligible order is explicitly indicative of intelligence.

Intelligence means only one thing, and it is not knowledge of "facts"; it is incorporeal Mind, What some still refer to as "Spirit". To use mind to deny Mind is the greatest absurdity a human being is capable of, scientist or not. Not "God", or "gods", or of any "supreme being(s)", but Mind as the intelligence that guides an apparently "random" action of Energy in an ultimate evolutionary direction (*ex + volvere*) as an "unrolling" or "rolling out"; and apart from all the add-ons to the term "entropy", entropy is instead as spontaneous change, transformation or turning of systems. What then, is the difference between "evolution" and "entropy", and of "Energy" as self-acting action guided by intelligent incorporeal Mind? The EWF model should then be called The Mind-Energy Wave Field.

13—Spacetime and the Mind-Energy Wave Field:

In Maxwell's electromagnetic (EM) theory, waves are generated by a "moving charge" which generates and moves through its own EM field. Might it rather be said that *a charge moves through spacetime, which generates an EM field?*

Einstein changed gravity from a special "force" to the "architectural geometry" of spacetime itself. For him, it was the presence of matter-mass that curved spacetime so that light must follow geodesic curvatures as the shortest distance between two points.

But what is spacetime? Space without content cannot be imagined much less measured, and *Time* without content cannot be measured as change since there is nothing that changes. Mind without content is Void and cannot be imagined at all without the imaginer there as content. So a question arises "*Is spacetime naught but Mind Itself?*"

In the Energy Wave Field (EWF) hypothesis, the unmanifest EW Field is *spacetime itself without content* except for the self-generated vibrations as a "wave field". *Energy* is nothing but self-acting action ie. vibrations generating unmanifest waves that, *Möbius-like*, twist and bend, folding back onto themselves to manifest as wave interference moiré configurations. But "matter" is nothing *but a direct effect* of Energy's self-acting and self-referring self-interferences. "Energy" as self-acting action is the essence of the word "dynamic".

Then what is *gravity*? If it is just the "architecture of spacetime", then there is no difference at all between "gravity-waves" and "Energy waves".

Thought and language divides and separates, and we assign different words to different things; but we also assign different words to unknowns, giving ourselves the air of somehow "knowing something". *Energy, spacetime and gravity* are three different words that refer to a "something" for which we have no real comprehension.

If the word "*Spirit*" refers to an unseen invisible "animating principle" (literally "the breath of life"), then that is closely related to, if not the same as the concept of Energy. The Christian idea of The Christ as *The Spirit made flesh* is analogous to *Energy manifest as matter*. Matter is Energy and nothing but Energy that is functioning in and as a specific wave pattern or form, and as form is a *created effect* having a beginning, it must also have an ending, whereas *Spirit/Energy* is *timelessly dynamic and eternal*. All thought, concepts and images are "form" and therefore like matter, are subject to Time as change and dissolution.

As thought is an *effect* of Mind in movement, of Mind twisting and bending back upon itself in *reflection*, thought is just a series of interference configurations, and as a measure of change, *is Time itself*. Thought *is* spacetime, gravity, matter and Energy. THEN WHAT IS IT THAT SEES ALL THIS?

END-NOTES

The numbers following each bold end-note refer to the page in this book. Topic is either in parens or underlined in reference. Cited reference-pages follow the reference site.)

1) p.—Amir Alexander, *Infinitesimal*.

2) p.—wikipedia.org/wiki/History_of energy, 2017, p. 2 of 4

3) p.-Manjit Kumar, *Quantum*, p. 22-25; Planck's "oscillators" were imaginary springs of varied tensions that interacted with EM radiant energy in indivisible "quantum packets", oscillating at fixed frequencies but at varying amplitudes. It occurs to me that Planck's imaginary springs were oscillating or spinning atoms and molecules of matter (which are the spheric wave-forms of this paper) interacting in the same ways as in photosynthesis (see "*Berkeley*," in text, pp and also fn.35, p.). I feel certain that it will turn out that EM radiation is not "quantized" as Einstein asserted, and that "*long term coherence*" of waves will be the rule rather than the exception, and that quanta after all will be the harmonic resonances between the pulsating frequencies of spheric matter wave forms and EM wave trains. "Oscillations"—whether of springs or particles— raises the question *"What makes an oscillator oscillate?"*

4) p.—(Salzburg) J.C. Polkinghorne, *The Quantum World*, p.54; Kumar, *Ibid.*, p. 63.

5) p.—Roger Penrose, *Cycles of Time, p.257* [2.3316] p.6.

6) p.—Baggott, *Farewell to Reality*, pp.55, 92-3; Muller, *Now—The Physics of Time*, pp.48, 56. "The *rest energy* is the intrinsic energy of an atom when at rest and not in motion. But a subatomic particle at rest is still the atom's

internal Ek as the oscillatory spin/rotation motion of its constituent protons, neutrons and electrons.—energy which is <u>not</u> intrinsic, but is still the "external" Ek motions and Ep positions of a body's constituent "particles".

7) p.—Regarding <u>sound waves</u>: All texts and authors I've read state that sounds waves are *longitudinal*; but sound waves generated in open air from a source of disturbance *radiate outward by compression and rarefaction* of the open-air molecules as shown. It is also maintained that sound waves are "legato" and not "staccato", but no vibration can be "legato"; a vibration is a series of rapid *pulses*, and a pulse cannot be "*legato.*"

8) p.—<u>Spheric waves</u> are not "oscillators"; a single spheric wave form will most likely *pulsate* outward, but when in interference with linear wave trains will result in <u>pairs of</u> <u>opposite pulsations</u>. Being *spheric,* they will likely have angular momentum (spin).

9) p.—"Acceleration" in physics means <u>any</u> change in speed (slowing or speeding up) or direction. In <u>refraction</u>, light waves moving obliquely from one medium's density into another is both a *change of speed and direction*, being therefore another way that a material medium affects the trajectory and speed of light; in Einstein's general relativity, acceleration is equivalent to gravity. Newton's concern with refraction of straight linear waves was probably because waves that are parallel to the density-change interface would exhibit no *bending* from one density to another; waves would continue parallel just as before, with only a change of wavelengh and frequency without any directional change. Only when waves are *oblique* to interface will there be any change of direction as bending as shown in figure 10 A and B.

10) p. —I can think of no experiment to determine the difference between an *optical* phenomenon and *gravitational bending* of light, but physicists may be able to devise such a test.

11) p.—https://physics.stackexchange. com/questions270982/how-does-the-light-source_fire-a-<u>single-photon</u>...? p.1 of 1. The question has been asked many times. The answer is it cannot.

12) p.—In quantum physics, the terms "<u>superposition</u>" and "<u>interference</u>" are more or less synonymous with "*<u>wave function.</u>*" In this paper, interference patterns are indeed effected by literally superposing two or more wave sets in intersection, but I suspect that in Q-physics, those are mostly if

not always, waves that are traveling "in line" with each other; *This paper emphasizes the superpositions intersecting on the diagonal between >0 and <90 degrees.* "Diffraction" in experiment seems confined to the two-slit effect where the waves do tend to spread out from a point source. But it is shown in figure 11 that secondary waves in oblique interferences do not "spread out" at all. However, *their wavelengths* are generally longer than the initial wave sets, but that is not "spread out" as Einstein meant.

13) p.—Weatherall, *Void,*, p.78; "The Electro-Magnetic field counts as "matter" in general relativity. Various configurations...have different amounts of energy and momentum associated, both of which influence curvature of space-time." And on page 125; "The geometry of space-time is rich and dynamic, *even without any matter at all.*" [italic mine] Burton Feldman, *Einstein's Genius Club*, p.34 "Energy is equivalent to mass; therefore it too can produce a gravitational curvature..."; Jim Baggott, *op cit.*, p.92: "Mass is a measure of a body's energy content.", and on p. 97 "...material object-masses are unnecessary for energy mass curvature."

14) p.—Anything can be in superposition, not just waves. The superposition of all possible states is the *wave-function*, which when an observation or measurment is made "collapses" from a wave to a "particle", or to the one state observed. This is also called "decoherence".

15) p.—(complexes) Matter as atoms and molecules would be more struc-tured than the figures show, and is better exemplified by the Platinum "diffraction" in figure 16.

16) p—(platinum) I've had this figure in my files for over thirty years, and have no idea where it was originally published; in this case, I absolve my-self from all liability charges of "intent to plagiarize" The original caption, as I indicated in the text, is incorrect, but there is no way of telling if that was the writer-publisher's doing, or if they were misled by some physi-cist's claim.

17) p.—interpretations of measurements and equations are usually dictated by the current theory, backed by an even less conscious metaphysical pre-supposition.

18) p.—Fulvio Melia, *Cracking the Einstein Code,* p.116; "Quanta such as photons may bubble up spontaneously out of the vacuum if an adequate source of energy lies nearby, *but they always form in pairs*...." [emphasis

mine] But the "vacuum" in the Energy Field.

19) p.—It will be shown in the text on page 52, that not only are <u>spheric waves produced from</u> <u>linear waves</u> in self-interference, but linears are also produced from spherics in self interference, which seems to make spherics and linears equivalent insofar as each can produce the other.

20) p.—In quantum theory, "either-or" is replaced by "and both" (or even "all"). All possible outcomes are "superposed" together into a multiple-component "wave-function". When a measurement-observation is made, the wave-function is said to "collapse" to a specific measurement observed. "Superposition" is "interference" and a hypothetical wave "collapses" to a "particle" in Q-theory. This "collapse of the wave function" is also referred to as "decoherence"; but it is noted in Baggott, *op cit.*, p. 212, that *"...decoherence is an assumption; we have no direct observational evidence that it happens"* [italics mine]. Mathematical equations are subject to interpretation, as well as the meaning of "measurement"; hence, the different "schools" of quantum theory, and different "universes" derived from Einstein's equations.

21) p.—("as though") Jim Baggott *ibid.*, p.30—Einstein believed that "photons are first and foremost particles...[whose] trajectories through space are predetermined by some property that leads to the *appearance of wave behavior as a result of statistical averaging*. [emphasis mine]; Kumar, *op cit.* pp.139-140, 186; Weatherall, *op cit.*, p.47,109; In March of 1905, Einstein "argued that for some purposes, light acts *as if* it is composed of what he called 'light quanta'—discrete, particle-like packets of energy." [my emphasis].

22) p.— Feldman, *op cit*, pp. 129, 134-5. Einstein knew of Newton's "corpuscular theory of light", and his own interpretation of "Brownian motion" as "random richochets" of molecules off of each other and off the sides of a container, influenced his "<u>light-particle</u>" idea. See also Baggott, *op cit,* p.31 and Kumar, *op cit.,* pp. 46,48.

23) p.—Baggott, ibid., p.30; Baggott attributes the "photon" name to the American chemist Gilbert Lewis in 1926 instead of to Leonard Trollard in 1916. With my dinosaur desktop and obsolete software, I have no luck finding any reference for Trollard's having named the "photon" in 1916 as stated in the text. Everyone uses the term "photon" as though it's a *fait*

accomplait, yet so few seem to reference the one who coined it.

24) p.—(non-acceptance) Baggott, *Ibid.*, p.29. Feldman, *op cit.,* p.131-32. Kumar, pp.33,46,96.

25) p.—(spreading waves) Feldman, *ibid..*, pp.131-2; Richard Muller, *Now,* p 118;. Polkinghorne, *op cit.*, p.7,131. Einstein also argued that increasing the *intensity* of light would *not* increase the number of electrons ejected but *would increase* their energy in the photo-electric effect, *which are both contradictory to what was actually observed and measured.*

26) p.—(intensity-frequency) Kumar, *op cit.*, pp.49-50.

27) p.—Since an electron particle's position—being a "cloud of probability"—is indeterminate, Hydrogen's electron-cloud can have many different different wave configurations.

28) p.—The "continuous wave" is often contrasted with the "discontinuous particle", but the fact that wave crests are assigned a "+" value (and the troughs "—") raises the question of what part of a wave is the "efficient" part? Using mathematical rules of signs as applied to nature seems arbitrary and oftentimes give results that boggle the rational mind, such as direction as + or the wave crest as "plus" and the trough as "minus" when the wave is asserted to be "continuous". The *wavelength* is a discrete measure, and the *frequency* is a discrete number. Crests and troughs can only be visualized as transverse, but a "top-down" or "aerial" view (as in the figures) can only show lines as crests and empty intervals as troughs. Also, contrary to Einstein, waves do not always "spread out" as they do in the diffraction figure. 11, in oblique superposed interferences, the waves do not spread at all in propagation.

29) p.—(string) This superposed wave configuration struck me almost as hard as the spheric wave pairs initially did. So many pieces seem to come together going back to Maxwell's original wave theory by *adding* quantum theory to waves as "*harmonic resonances*", but NOT as particles. This interference "string" configuration also gets rid of all the extra super-tiny invisible dimensions that makes string theory untestable and untenable.

30) p.—Kumar, *Ibid.*, pp. 139-140.

31) p —The "photon" is weird; being "massless" (ie. having zero "rest mass") it can never _not_ be in motion, therefore it *must have* "momentum", despite having no mass *other than* its kinetic energy (Ek). But because of the

equivalence of energy and mass (E=mc²), being "mass-less", it cannot <u>be energy</u>, and is "demoted" to be a mere "*carrier* of energy" at last is even suggested to be "*guided by a pilot-wave*". I say, "Snuff the photon!" This exemplifies the habit of relying on a series of *ad hoc* adjustments to uphold an increasingly shaky theory.

32) p.—(photon momentum) Baggott, *op cit.*, pp.27, 32-3;

33) p..—(pilot) Baggott, *ibid.*, pp.210-11;Kumar, op cit., pp.335-6; Polkinghorne, *Ibid.*, p. 57.

34) p.—(light box) Kumar, *ibid.*, pp.282-7.

35) p. (laser) The "stimulating lamp" is probably not correct as shown in the illustration, since I have no accurate diagrams; suffice it to say that there is some kind of "external light" to stimulate EM energy emission from each atom's electron(s that is then amplified back-and-forth by reflections, gathering momentum to burst out of the crystal's less-silvered front end.

36) p.—Deepak Chopra and Menas Kafatos, *You Are The Universe*, pp.177-8. Baggott, op cit., pp. 36 .I suspect that in the near future more evidence will be accumulating against the "particle", "super-position" as a "packet" of alternate outcomes and the "decoherence/collapse of the wave-function" from a "probability-wave-to-a-particle". The "quantum" will be upheld as something like a *"harmonic multiple of some fundamental frequency"* as a wave-packet, but a spheric wave might be treated mathematically as a "particle", but not as a *point*.

37) p.—Paul A. Laviolette, *Decoding the Message of the Pulsars*, p. 59;...*synchrotron radiation from our Galactic core is mainly <u>circularly polarized</u>.* [The author cites fn 6: *Bower, Falcke and Backer, "Circular polarization in Sagittarius A",* at 195th *American Astronomical Society* meeting, Atlanta, January., 2000]. "... our Galactic core is mainly <u>circularly polarized</u>...produced only when cosmic rays travel toward the observer at dose to the speed of light, following a spiral trajectory."* [All italics are the author's; underlines are mine]; Muller, *op cit.*, p.220 fn: "Real 3D glasses typically...use "circular" polarization, which makes the effect insensitive to the angle of the viewing human head."[italic mine]; Stuart Clark, *The Unknown Universe*, p. 233: The gravitational waves in the [CMB] were inferred because of the detection of...*circular polarization...a <u>corkscrewing motion in the microwaves</u> implanted* by the passage of gravitational

waves."[italics and underline mine]. I believe that since all EM energy-radiating bodies rotate (*ie. "spin"*)—that must impart a *twist* to emitted radiation, which would therefore travel in a spiraling trajectory. Somehow gravity is connected to spin, circular polarization, and synchrotron radiation, and also to the problem of *shear* discussed in Melia, *op cit.*, pp,43-47. That shear problem (in calculating Einstein's equations) was "solved" in this manner: (p.47); "Why bother making the shear drop off rapidly with distance? Why not just consider shear-free geodesics everywhere?" [Melia citing Robinson and Trautman *in text, not in fn.*]. Cutting a pair of shears along the edge of a sheet of paper makes the cut edge coil into a spiral. It is suggested that instead of eliminating shear as reported in Melia, *that shear is significant* in the interactions between gravity and EM radiation, *and can produce a spiraling motion to EM propoagation even without synchrotronic black hole radiation.*

38) p.—All <u>mathematic</u> rules and number theory are from Friedrich Waismann, *Introduction to mathematical Thinking*, Harper and Row, 1951.

39) p.—(infinities) William Wallace, *The Logic of Hegel*, pp.174-178.

40) p.—(Gödel) Feldman, *op cit.*, pp.89-90; Muller, op cit., pp.257-58; Waismann, op cit. pp.100, 217.

41) p.—(quantum) Penrose, *op cit.*, pp.185, 204.

42) p.—(Doppler) *ibid.*, p.60. In Roger Penrose' Cycles of Time, page 60, figure 2-1 shows an increased red-shift between a "near star" and a far "distant galaxy"; however, this only shows a red-shift increase over distance and not necessarily over time as expansion-recession. To indicate Doppler recession, the red-shift for a single object would have to be measured over enough span of time to show recession. Since the hypothesis is that universe is expanding at a faster and faster rate of increased recession, a six month's or a year's interval should be sufficient to show any increase of red shift as Doppler.

43) p.—(Zeeman) Feldman, *op cit.*, p103-106; Kumar, op cit., p. 386.

44) p.—(Zwicky) Clark, op cit..,p.181; Charles Seife, *Alpha & Omega*, pp.225-227.

45) p.—There are some who argue that the universe's rate of expansion is not restricted by <u>light-speed</u> and that "c" *only* pertains to EM energy.

46) p.—Penrose—op cit.—discusses <u>entropy</u> at length over many pages in relation to "cycles"; *but on pages 40-41, he cites an experiment where*

entropy seems to actually reverse. The "second law of thermodynamics" is not thoroughly clarified, especially as connected to time's one-dimensional movement. The Greek *en* = "in" + *trope* means only "in transformation"—change—but does not say anything about change as *direction toward increasing disorder.*

47) p.—It is believed for all spiral galaxies (and maybe barred spirals too), that at the very center is a super mass-density black hole that effects a slowing of time, accompanied by a red-shift for those observers outside the gravitational well. It is supposed at the Schwarzchild radius or "event horizon", that time for an "outside" observer would appear to slow or even come to a stop (Penrose op cit., pp.96-97; Scharf, *op cit.*, pp.26-8, 79-80, 107-108: Muller, *op cit.*,pp.83, 87, 164-166). There would be no greater red-shift than that from the many black-hole galaxies in the interval between an observer and deep space observations. Peering into deep space, these gravitational effects would *collectively* and cumulatively effect extremely high degrees of red-shift; assuming this to be due to the Doppler effect would inevitably be interpreted as accelerating recession, followed by the logical supposition that the recession-expansion was due to a big bang explosive beginning—and it would be wrong. I doubt that this collective black hole gravitational effect has been considered relative to the cosmic red-shift Doppler assumption, but the whole big bang hinges entirely on the Doppler assumption. I've always been suspicious of the assumed connection between two independent observations—the cosmic red-shift, and the Doppler effect and I do believe that the "evidence" (Penrose, *Cycles of Time*, p.60) can be interpreted differently as gravitational red-shift from galactic black holes. So the question is, *Is the universe really expanding at an ever increasing rate or not?* It is the position in this paper that The EWF universe is considered to be in a *dynamic equilibrium* with variable mass-densities and entropy increasing and *decreasing* in different regions or systems, *but constant overall,* and with gravitational curvatures that effect rapid expansion "here", and extremely slow "regional" expansions or even contractions "there" in a universe that is eternally "turning itself inside-out and outside-in" in dynamic equilibrium.

48) p.—(LQG) *The Week* magazine, 1 February, 2013 in "Science" section.

49) p.—(Void) *The Week magazine*, 15 May, 2015, in "Science" section.

50 p.-(LOS) Even though EM energy is emitted from stars as *radiant* waves, and those waves are all around us all the time, it is only by direct line-of-sight (LOS) contact with the eye's retina as well as with any instruments that emitted or reflected light waves are observable or measurable, and only those frequencies that the retina is inherently sensitive to, which is why we see stars as "pin-points" during direct loss.

51) p.—"Energy" as "work" or "action" has never been observed, measured, touched or detected by senses or instruments. The invisible EWF is itself "dark", and is only *inferred* from observing its manifest effects as radiation, gravity, and the behavioral effects of matter, . Positing an unknowable "dark energy" in addition to "normal" but unknown energy as an attempt to explain the increasing rate of expansion is no better than positing "God", "Satan", Darth Vader or leprechauns as the "dark doers".

52) p.—Wikipedia, "zero-point energy"— Vacuum birefringence. p.3 of 6.

53) p.—(variables) Clark, *op cit.*, pp. 258-260, By computing energy-mass densities in different regions of the universe, *Thomas Buchert* finds *that the universe is not expanding equally*. The universe has a huge near-void in one major region with rapidly accelerating expansion, and a high energy-mass density of high gravitational pull in another major region that is rich in galactic clusters that expands much more slowly. "*It's inevitable that time will pass at different rates in different parts of universe too...there is no single age of the universe.*" Only after making these regional measurements does Buchert do an average from their sum.

54) p.—(adiabatic heat) "Heat" in physics refers to heat-flow from a higher into a lower temperature which—in finite systems—can reach equilibrium, so adiabatic is not as superfluous as I maintain in the text. However, the *universe* is designated by cosmological physicists to be an "adiabatic system" *in order for heat to not be able to cross the boundary, thus trapping it* so that "waste heat" will accumulate entropically and the universe dies a hot but static end. The claim is that "energy is "present (ie. to obey the first law) but is "*unavailable to do any work*". That's where I see the *ad hoc* subterfuge to accommodate the space-time expansion model based on Doppler red-shift, despite it being in opposition to the law of energy conservation. See Penrose, *op cit.*, p. 72.

55) p.—(CMB) Baggott, *op cit*, pp.112-115,121; Penrose, *op cit.*, pp. 69-71, 216-219; Wallace Tucker, *Chandra's Cosmos*, pp.51-53.

56) p.—(Auriga *Galaxy Cluster 4C41.17* & CMB) Caleb Scharf, *Gravity's Engines*, pp. 4, 8, 154-155, 160.

57) p —Moving the axis of the spinning spheres is obviously equivalent to the observer changing the reference frame (ie. the angle of view), and the only spin that is <u>observer IN-dependent</u> is L-R and R-L on the vertical axis.

58) p.—The <u>puzzle</u> is that for observers in front and in back of the vertical axis spin, either observed spin is *observer independent*; but still it's just weird that *that same spin* observed from either of the horizontal axes' spins' *"becomes" observer dependent*! One can reason it based on our vertical-axis bilateral symmetry—but it's still strange.

59) p.—(Gödel) Feldman, *op cit.*, p.90.

60) p.—See text page 25, Fn 36 for "*corkscrew-spiral*" propagation references.

61) p.—This information was notated from a *Great Courses* course (number 1830) on "cosmology" in which I found the instructor pompous, and have since given the DVD and booklet away unfortunately without noting the instructor's name. But the 2% margin of error nagged at me, so with a little calculating (not my forte) I saw how much the nagging was. If there are any miscalculations or arithmetic errors, they are mine, not the instructor's.

62) p.—*The Seattle Times*, 26 Feb,2017, "News", A8.

63) p.—Noson S.Yanofsky, *The Outer Limits of Reason*, p.172.

64) p.—Unlike "normal matter", "dark matter" appears to not interact with EM energy, nor can it be detected by any other means; it is strictly inferred to "explain" the actions of "normal matter" whose behavior is anomalous within the "standard model predictions.

65) p.—http://wikipedia.org/wiki/Gravity, 6/16/17, "Anomalies and discrepancies", p.9 of 14.

66) p.—Melia, *op cit.,* p.34; Muller, *op cit.*, p.74; LaViolette, *op cit.*, p.46-7; Weatherall, *op cit.*,pp.63-04,71-2,78-9,125-6, and "DeSitter universe" on p.76.

67) p.—This is another way that matter can cause Light to "accelerate", effecting more "local" gravitational curvatures *after* the initial Energy-mass density curvature (which is antecedent to spheric matter waves produced by Energy wave interferences).

68) p.—The Energy of the unmanifest and immeasurable EWF is completely different from the "energy" of thermal equilibrium" in that the latter is

"finished" with *no potential for further action, whereas the former is nothing but pure potential for self-interfering action.*

69) p.—It is to be noted that since the manifest EM energy seems to propagate in a "corkscrewing, twisting" motion, that it is no stretch to imagine un-manifest Energy waves twisting Möbius-like as space-time curvature *prior* to long-term interferences that produce the spheric matter-wave forms. This "twist-and bend" allows the field of energy waves to self-interfere as shown in this paper, and also imparts shear, circular polarization and *spiral-spin* to all spheroidal bodies, subatomic and cosmic.

70) p.—It's already been shown in fn 45 above that there seem to be exceptions to this "rule"

71) p.—Adam Frank, *About Time*, pp.254, 267.

72) p.—Penrose, *op cit.*, pp.103,124.

73) p.—I sometimes wonder how much the Judeo-Christian *Genesis* plays into the "Big Bang" creation myth.

74) p.—Weathrall, *op cit.,* pp .9,77, 112-113, 125, 134.

75) p.—The "calculators' emphasis" on the problem of "information and entropy" is a result of the focus on "data collection" instead of intelligence (NOT military or agencies in which "intelligence" is an oxymoron). "In-formation" is just another *form* no different than spheric *matter-wave forms*. All created form and pattern is subject to entropic dispersion and dissolution.

76) p.—Eliot Deutsch, *Advaita Vedanta*, p.31, (fn. 8).

77) p.—Feldman, *op cit.,* p.196 (Einstein quote)

78) p.—The real point is that whatever the form, a "primary" wave-form is the fundamental self-interfering *source* for effecting both linear wave trains and spheric wave forms that produce all successive linears and all spheric *wave pairs—and all phenomenal forms and processes as fractal iterations of the self-referential original.*

79) p.—Baggott, *op cit.*, p.145.

80) p.—Scharf, op.cit., p.119.

BIBLIOGRAPHY

(asterisk* indicates sources used in references)

Alexander, Amir,* *Infinitesimal*, Scientific American / Farrar, Straus, Giroux, NY., 2014.

Baggott, Jim,* *Farewell to Reality*, Pegasus Books LLC., NY., 2013.

Chopra, Deepak & Kafatos, Menas,* *You Are the Universe*. Harmony/Crown, NY. 2017.

Clark, Stuart,* *The Unkown Universe*, Pegasus Books LTD., NY., 2016.

Deutsch, Eliot,* *Advaita Vedanta*, East-West Center Press, Honolulu, 1969.

Feldman, Burton,* *Einstein's Genius Club*, Arcade, NY., 2011.

Frank, Adam,* *About Time*, Free Press, NY., 2011.

Gleick, James, *Chaos*. Penguin, NY., 1987.

Godfrey-Smith, Peter, *Other Minds*, Farrar, Straus, Giroux, NY., 2016.

Kerrod, Robin, and Carole Scott, *Hubble*, Firefly Books LTD. Canada, 2011.

Kotakowski, Laszeck, *Metaphysical Horror*, Univ. of Chicago Press / Penguin, 2001.

Krishnamurti, J. and Dr. David Bohm, *Ending of Time*, Harper and Row, SF., CA., 1985.

Krishnamurti, J., *The Awakening of Intelligence*, (Dialogues with Dr. David Bohm), Harper and Row, London, UK.,1973.

Krishnamurti, J., *Truth and Actuality*, (Dialogue with Dr. David Bohm), Krishnamurti Foundation, London, 1980.

Kumar, Manjit,* *Quantum*, Omnibus, London, UK., 2009.

Livio, Mario, *The Golden Ratio*, Broadway, Books, NY., 2002.

Melia, Fulvio,* *Cracking the Einstein Code*, University of Chicago Press, 2009.

Miller, A.F. (tr.), *Hegel's Philosophy of Mind*. Oxford-Clarendon, London, UK., 1971.

Muller, Richard A.,* *Now; The Physics of Time*, W.W.Norton and Co., NYC., 2016.

Penrose, Roger,* *Cycles of Time*, Alfred A. Knopf, NY., 2010.

Peters, F.E., *Greek Philosophical Terms*, New York University Press, 1967.

Pickover, Clifford, *The Möbius Strip*, Avalon Group, NY., 2006.

Polkinghorne, J.,* *The Quantum World*, Princeton University Press, NJ., 1984

Scharf, Caleb,* *Gravity's Engines,* Scientific American/Farrar, Straus, Giroux, NY. 2012.

Schwartz, Jeffrey, *Sudden Origins*, John Wiley and Sons, NY., 1999.

Seife, Charles,* *Alpha and Omega*, Viking, NY., 2003.

Stewart, Ian, *Calculating the Cosmos*, Basic Books / boat Enterprises, 2016.

Tomlin, E.W.F., *The Oriental Philosphers*, Harper and Row, NY., 1963.

Tucker, Wallace, H.,* *Chandra's Cosmos*, Smithsonian, WA.DC., 2017.

Waismann, Fredrich* *Introduction to Mathematical Thinking*, Harper and Row, NY. 1951

Wallace, William, (tr.),* *The Logic of Hegel*, Oxford University Press, London, UK., 1965.

Weatherall, James Owen,* *Void,* Yale University Press, London, UK., 2016.

Wilbur, Ken (ed.), *The Holographic Paradigm*, New Science Library, NY., 1982.

Wolfson, Richard, *Physics and our Universe*, The Teaching Company, VA., 2011.

Yanofsky, Noson,* *The Outer Limits of Reason*, M.I.T. Press, 2013.

Periodicals

*The Seattle Times** newspaper, 26 February, 2017 in "News" sect. A, pg. 8

*The Week** magazine, 1 February 2013 in "Science"; 15 may, 2015, in "Science

Internet

http://physics.stackexchange.com/questions270982/how-does-the- l i g h t - source-fire-a-

single-photon...? p.1 of 1.*

http://astro.wku.edu.astr106/Hubble_intro.html. "Hubble's Distance-Redshift Relation". 6 pgs. 2/2/18.

http://electric-cosmic.org/arp/htm "Halton Arp's discoveries about redshift", 3 pgs, 3/1/17. http://www.cs.mcgill.ca/ rwest/wikispeedia/wpcd/r/Red-shift.htm.

(NOTE; The following all begin with http://wikipedia.org/wiki/ *followed by the specific site;* brackets [ ] are not part of the site address, but indicate end of the last, and start of the next site)

Black-body_radiation, 8/10/17, 24 pgs.] [Casimir_effect, 12/11/16, 4 pgs. (vacuum energy, p.3.

Van der Waals, p.4)] [Cosmological_principle, 3/15/17, 5 pgs.] [De-Sitter-un-iverse, 3/23/17, 3 pgs.]

[Energy, 1/13/19, 19 pgs.,(History, p. 4; Conservation, pp.11-12; Thermo-dynamics, pp15-16]

[Entropy,___27 pgs.] [Gravity, 6/16/17, 14 pgs. (Anomalies, p.9)] [Gravita-tional redshift, 7/24/17, (General Relativity Redshift (observed) vs. de-duced Gravitational Time Dilation, p.7)]

[Gravitational_time-dilation, 7/24/17, 6 pgs] [Harmonic_oscillator, 7/15/17, 14 pgs]

[History_of energy, 2/16.17, 4 pgs. ("vis viva", p.1; thernodynamics, p.2.)-1 [History_of entropy, 5/22/17, 8 pgs.]

[Infinity, 6/18/17, 13 pgs.] [Isaac Newton, 6/10/17] [Joseph von Fraunhofer, 3/13/17, 3 pgs.]

[Mind, 6/19/17, 16 pgs.} [Mixmaster, ___3 pgs. (misner, p.1)] [Emmy Noether, /3/2/17]

[Photon, 8/6/17, 6 pgs.] [Photon_gas, 8/7.17, 5 pgs.] [Planck epoch, 1/24/17,

18 pgs. (with Planck units)]

[Relativistic_Doppler_effect, 2/28/17, 12 pgs.] [Schrödinger_equation, 6/24/17, 33 pgs.-]

[spectroscopy, 3/26/17, 9 pgs.] [Stellar_rotation, 2/27/17/, 8 pgs.] [Tensor, 5/17/17/, 17 pgs:]

[Vacuum_energy, 3/22/17, 5 pgs.] [Van_der Waals force, 3/26/17, 7 pgs.] [Zero-point energy, 6/14/17, 10 pgs. (vacuum birefringence, p.6)] (<u>end of Internet site cites</u>).

INDEX

PROPER NAMES

M

manifest (EWF)—41-42, 44, 49, 58

many worlds—1-2, 54

mass-9, 44, 66, 69

    density—23, 69

        differential curvature—xxvi, 23, 44, 69

    and energy—81fn13

    inertia—67

84fn28

    momentum—67

    rest mass—

mathematics—13, 16, 25, 65(1), 85fn38

    calculus—16

    rules and reality—16, 65, 85fn38

matter—xi, xiii, 15, 30, ,44, 66, 77

    anti-matter—xxix

        complexes—vi, 75

        dark—22, 24, 36, 71(6), 88fn64

        and Energy—xiii, 43, 62, 77(13)

measure –xiv, 23-4, 48

metaphysics and physics—16, 57

memory—50-3

    universal—52-3

mind—49, 61-2

Mind (Intelligence)—xvi, 3, 59-60, 62-3, 76-7(13)

    and Energy—xvi, 59, 63, 76-7(13)

    Geist—62

    Hsin—62

    self-referencing—44-5, 49, 60, 89fn78

Mind-Energy Wave Field—76-7(13)

moiré—v, 42, 48, 61

Möbius twist—xiv, 13, 39, 63, 71(5F), 77(13)

    shear—xxx, 71(5F), 88fn60

    signature—15, 33-4, 65, 83fn28

momentum—51, 67

motion and position—xii, 34-5, 42-3

    and LOS—23, 31, 87fn50

N

Newton's trick—33

Number System—13-17

    continuous-discrete—16, 83-

    Godel—16

    Imaginary—15, 17

    Real—13

    signature—15, 33-4, 65, 83fn28

O

observer-observed –41-3, 52, 56-7

    dependent/independent—28, 52

    in quantum theories—52

order—57-9

oscillations –xiii, 5, 13, 30, 49, 73fn3

P

parallax—34

particle—xi, 3

    Brownian motion—29

    corpuscle—xii

    discrete, and continuous—16, 83fn28

pattern—46-4, 62

perception—54-5

periodic numbers—14

    waves—xvii

photo-electric effect—xiii, 2, 5

photosynthesis—vi, 11-12